Electroceramics: Materials, Properties and Applied Principles

Electroceramics: Materials, Properties and Applied Principles

Edited by
Quentin Merton

WILLFORD PRESS
www.willfordpress.com

Published by Willford Press,
118-35 Queens Blvd., Suite 400,
Forest Hills, NY 11375, USA

ISBN: 978-1-68285-488-4

Cataloging-in-Publication Data

Electroceramics : materials, properties and applied principles / edited by Quentin Merton.
   p. cm.
Includes bibliographical references and index.
ISBN 978-1-68285-488-4
1. Electronic ceramics. 2. Electronics--Materials. 3. Electronic ceramics--Chemistry.
I. Merton, Quentin.
TK7871.15.C4 E44 2018
621.381--dc23

For information on all Willford Press publications
visit our website at www.willfordpress.com

WILLFORD PRESS

Contents

Permissions

Index

Preface

Any ceramic material, which is used for its electrical properties is known as electroceramics. These electroceramics are the materials that are used for their distinct storage, magnetic and optical properties. The different forms of electroceramics are fast ion conductor ceramics, magnetic ceramics, dielectric ceramics, electronically conductive ceramics and piezoelectric and ferroelectric ceramics. This book unfolds the innovative aspects of electroceramics, which will be crucial for the holistic understanding of the subject matter. As this field is emerging at a rapid pace, the contents of this textbook will help the readers understand the modern concepts and applications of the field.

A detailed account of the significant topics covered in this book is provided below:

Chapter 1- Electroceramics are a kind of ceramics that have unique electrical properties. They include dielectric, semiconducting, ionically conducting and superconducting materials. This is an introductory chapter which will introduce briefly all the significant aspects of electroceramics.

Chapter 2- Magnetic ceramics possess strong magnetic coupling, low loss characteristics and high electrical resistivity. They are used in data storage, information transmission, power supply, etc. This chapter has been carefully written to provide an easy understanding of the varied facets of magnetic ceramics.

Chapter 3- Ferroic ceramics are materials that exhibit ferromagnetic, ferroelastic or ferroelectric properties. A hysteresis effect is required for such effects to take place. They can be used to make magnetoelectric devices. Electroceramics is best understood in confluence with the major topics listed in the following chapter.

Chapter 4- Dielectrics are ceramic materials that do not conduct electricity. They are primarily used in the development of capacitors and resonators. The aspects elucidated in this chapter are of vital importance, and provide a better understanding of electroceramics.

I would like to make a special mention of my publisher who considered me worthy of this opportunity and also supported me throughout the process. I would also like to thank the editing team at the back-end who extended their help whenever required.

Editor

An Introduction to Electroceramics

Electroceramics are a kind of ceramics that have unique electrical properties. They include dielectric, semiconducting, ionically conducting and superconducting materials. This is an introductory chapter which will introduce briefly all the significant aspects of electroceramics.

Electroceramics

Electroceramics is a class of ceramic materials used primarily for their electrical properties.

While ceramics have traditionally been admired and used for their mechanical, thermal and chemical stability, their unique electrical, optical and magnetic properties have become of increasing importance in many key technologies including communications, energy conversion and storage, electronics and automation. Such materials are now classified under *electroceramics*, as distinguished from other functional ceramics such as advanced structural ceramics.

Historically, developments in the various subclasses of electroceramics have paralleled the growth of new technologies. Examples include: ferroelectrics - high dielectric capacitors, non-volatile memories; ferrites - data and information storage; solid electrolytes - energy storage and conversion; piezoelectrics - sonar; semiconducting oxides - environmental monitoring.

Dielectric Ceramics

Dielectric materials used for construction of ceramic capacitors include: zirconium barium titanate, strontium titanate (ST), calcium titanate (CT), magnesium titanate (MT), calcium magnesium titanate (CMT), zinc titanate (ZT), lanthanum titanate (TLT), and neodymium titanate (TNT), barium zirconate (BZ), calcium zirconate (CZ), lead magnesium niobate (PMN), lead zinc niobate (PZN), lithium niobate (LN), barium stannate (BS), calcium stannate (CS), magnesium aluminium silicate, magnesium silicate, barium tantalate, titanium dioxide, niobium oxide, zirconia, silica, sapphire, beryllium oxide, and zirconium tin titanate Some piezoelectric materials can be used as well; the EIA Class 2 dielectrics are based on mixtures rich on barium titanate. In turn, EIA Class 1 dielectrics contain little or no barium titanate.

Electronically Conductive Ceramics

Indium tin oxide (ITO), lanthanum-doped strontium titanate (SLT), yttrium-doped strontium titanate (SYT).

Fast Ion Conductor Ceramics

Yttria-stabilized zirconia (YSZ), gadolinium-doped ceria (GDC), lanthanum strontium gallate magnesite(LSGM), beta alumina, beta" alumina.

Piezoelectric and Ferroelectric Ceramics

Commercially used piezoceramic is primarily lead zirconate titanate (PZT). Barium titanate (BT), strontium titanate (ST), quartz, and others are also used.

Magnetic Ceramics

Ferrites including Iron(III) Oxide and Strontium carbonate display magnetic properties. Lanthanum strontium manganite exhibits colossal magnetoresistance.

Ceramic

A selection of silicon nitride components.

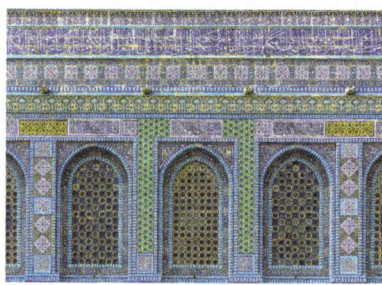

Mid-16th century ceramic tilework on the Dome of the Rock, Jerusalem

Spherical Hanging Ornament, 1575-1585, Ottoman Period. Brooklyn Museum.

A ceramic is an inorganic, non-metallic, solid material comprising metal, non-metal or metalloid atoms primarily held in ionic and covalent bonds. This article gives an overview of ceramic materials from the point of view of materials science.

The crystallinity of ceramic materials ranges from highly oriented to semi-crystalline, vitrified, and often completely amorphous (e.g., glasses). Most often, fired ceramics are either vitrified or semi-vitrified as is the case with earthenware, stoneware, and porcelain. Varying crystallinity and electron consumption in the ionic and covalent bonds cause most ceramic materials to be good thermal and electrical insulators (extensively researched in ceramic engineering). With such a large range of possible options for the composition/structure of a ceramic (e.g. nearly all of the elements, nearly all types of bonding, and all levels of crystallinity), the breadth of the subject is vast, and identifiable attributes (e.g. hardness, toughness, electrical conductivity, etc.) are hard to specify for the group as a whole. General properties such as high melting temperature, high hardness, poor conductivity, high moduli of elasticity, chemical resistance and low ductility are the norm, with known exceptions to each of these rules (e.g. piezoelectric ceramics, glass transition temperature, superconductive ceramics, etc.). Many composites, such as fiberglass and carbon fiber, while containing ceramic materials, are not considered to be part of the ceramic family.

The earliest ceramics made by humans were pottery objects (i.e. *pots* or *vessels*) or figurines made from clay, either by itself or mixed with other materials like silica, hardened, sintered, in fire. Later ceramics were glazed and fired to create smooth, colored surfaces, decreasing porosity through the use of glassy, amorphous ceramic coatings on top of the crystalline ceramic substrates. Ceramics now include domestic, industrial and building products, as well as a wide range of ceramic art. In the 20th century, new ceramic materials were developed for use in advanced ceramic engineering, such as in semiconductors.

The word "*ceramic*" comes from the Greek word (*keramikos*), "of pottery" or "for pottery", from (*keramos*), "potter's clay, tile, pottery". The earliest known mention of the root "ceram-" is the Mycenaean Greek*ke-ra-me-we*, "workers of ceramics", written in Linear B syllabic script. The word "ceramic" may be used as an adjective to describe a material, product or process, or it may be used as a noun, either singular, or, more commonly, as the plural noun "ceramics".

Types of Ceramic Material

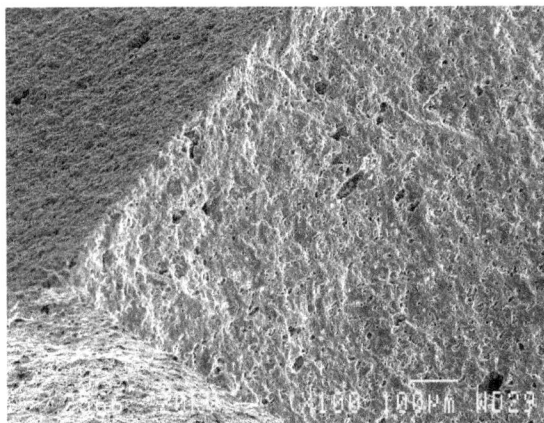

A low magnification SEM micrograph of an advanced ceramic material. The properties of ceramics make fracturing an important inspection method.

A ceramic material is an inorganic, non-metallic, often crystalline oxide, nitride or carbide material. Some elements, such as carbon or silicon, may be considered ceramics. Ceramic materials are brittle, hard, strong in compression, weak in shearing and tension. They withstand chemical erosion that occurs in other materials subjected to acidic or caustic environments. Ceramics generally can withstand very high temperatures, such as temperatures that range from 1,000 °C to 1,600 °C (1,800 °F to 3,000 °F). Glass is often not considered a ceramic because of its amorphous (noncrystalline) character. However, glassmaking involves several steps of the ceramic process and its mechanical properties are similar to ceramic materials.

Traditional ceramic raw materials include clay minerals such as kaolinite, whereas more recent materials include aluminium oxide, more commonly known as alumina. The modern ceramic materials, which are classified as advanced ceramics, include silicon carbide and tungsten carbide. Both are valued for their abrasion resistance, and hence find use in applications such as the wear plates of crushing equipment in mining operations. Advanced ceramics are also used in the medicine, electrical, electronics industries and body armor.

Crystalline Ceramics

Crystalline ceramic materials are not amenable to a great range of processing. Methods for dealing with them tend to fall into one of two categories – either make the ceramic in the desired shape, by reaction *in situ*, or by "forming" powders into the desired shape, and then sintering to form a solid body. Ceramic forming techniques include shaping by hand (sometimes including a rotation process called "throwing"), slip casting, tape casting (used for making very thin ceramic capacitors, e.g.), injection molding, dry pressing, and other variations. A few methods use a hybrid between the two approaches.

Noncrystalline Ceramics

Noncrystalline ceramics, being glass, tend to be formed from melts. The glass is shaped when either fully molten, by casting, or when in a state of toffee-like viscosity, by methods such as blowing into a mold. If later heat treatments cause this glass to become partly crystalline, the resulting material is known as a glass-ceramic, widely used as cook-top and also as a glass composite material for nuclear waste disposal.

Properties of Ceramics

The physical properties of any ceramic substance are a direct result of its crystalline structure and chemical composition. Solid state chemistry reveals the fundamental connection between microstructure and properties such as localized density variations, grain size distribution, type of porosity and second-phase content, which can all be correlated with ceramic properties such as mechanical strength σ by the Hall-Petch equation, hardness, toughness, dielectric constant, and the optical properties exhibited by transparent materials.

Physical properties of chemical compounds which provide evidence of chemical composition include odor, colour, volume, density (mass / volume), melting point, boiling point, heat capacity, physical form at room temperature (solid, liquid or gas), hardness, porosity, and index of refraction.

Ceramography is the art and science of preparation, examination and evaluation of ceramic microstructures. Evaluation and characterization of ceramic microstructures is often implemented on similar spatial scales to that used commonly in the emerging field of nanotechnology: from tens of angstroms (A) to tens of micrometers (μm). This is typically somewhere between the minimum wavelength of visible light and the resolution limit of the naked eye.

The microstructure includes most grains, secondary phases, grain boundaries, pores, micro-cracks, structural defects and hardness microindentions. Most bulk mechanical, optical, thermal, electrical and magnetic properties are significantly affected by the observed microstructure. The fabrication method and process conditions are generally indicated by the microstructure. The root cause of many ceramic failures is evident in the cleaved and polished microstructure. Physical properties which constitute the field of materials science and engineering include the following:

Mechanical Properties

Cutting disks made of silicon carbide

Mechanical properties are important in structural and building materials as well as textile fabrics. They include the many properties used to describe the strength of materials such as: elasticity / plasticity, tensile strength, compressive strength, shear strength, fracture toughness&ductility (low in brittle materials), and indentation hardness.

In modern materials science, fracture mechanics is an important tool in improving the mechanical performance of materials and components. It applies the physics of stress and strain, in particular the theories of elasticity and plasticity, to the microscopic crystallographic defects found in real materials in order to predict the macroscopic mechanical failure of bodies. Fractography is widely used with fracture mechanics to understand the causes of failures and also verify the theoretical failure predictions with real life failures.

The Porsche Carrera GT's carbon-ceramic (silicon carbide) disc brake

Ceramic materials are usually ionic or covalent bonded materials, and can be crystalline or amorphous. A material held together by either type of bond will tend to fracture before any plastic deformation takes place, which results in poor toughness in these materials. Additionally, because these materials tend to be porous, the pores and other microscopic imperfections act as stress concentrators, decreasing the toughness further, and reducing the tensile strength. These combine to give catastrophic failures, as opposed to the normally much more gentle failure modes of metals.

These materials do show plastic deformation. However, due to the rigid structure of the crystalline materials, there are very few available slip systems for dislocations to move, and so they deform very slowly. With the non-crystalline (glassy) materials, viscous flow is the dominant source of plastic deformation, and is also very slow. It is therefore neglected in many applications of ceramic materials.

To overcome the brittle behaviour, ceramic material development has introduced the class of ceramic matrix composite materials, in which ceramic fibers are embedded and with specific coatings are forming fiber bridges across any crack. This mechanism substantially increases the fracture toughness of such ceramics. The ceramic disc brakes are, for example using a ceramic matrix composite material manufactured with a specific process.

Electrical Properties

Semiconductors

Some ceramics are semiconductors. Most of these are transition metal oxides that are II-VI semiconductors, such as zinc oxide.

While there are prospects of mass-producing blue LEDs from zinc oxide, ceramicists are most interested in the electrical properties that show grain boundary effects.

One of the most widely used of these is the varistor. These are devices that exhibit the property that resistance drops sharply at a certain threshold voltage. Once the voltage across the device reaches the threshold, there is a breakdown of the electrical structure in the vicinity of the grain boundaries, which results in its electrical resistance dropping from several megohms down to a few hundred ohms. The major advantage of these is that they can dissipate a lot of energy, and they self-reset – after the voltage across the device drops below the threshold, its resistance returns to being high.

This makes them ideal for surge-protection applications; as there is control over the threshold voltage and energy tolerance, they find use in all sorts of applications. The best demonstration of their ability can be found in electrical substations, where they are employed to protect the infrastructure from lightning strikes. They have rapid response, are low maintenance, and do not appreciably degrade from use, making them virtually ideal devices for this application.

Semiconducting ceramics are also employed as gas sensors. When various gases are passed over a polycrystalline ceramic, its electrical resistance changes. With tuning to the possible gas mixtures, very inexpensive devices can be produced.

Superconductivity

The Meissner effect demonstrated by levitating a magnet above a cuprate superconductor, which is cooled by liquid nitrogen

Under some conditions, such as extremely low temperature, some ceramics exhibit high temperature superconductivity. The exact reason for this is not known, but there are two major families of superconducting ceramics.

Ferroelectricity and Supersets

Piezoelectricity, a link between electrical and mechanical response, is exhibited by a large number of ceramic materials, including the quartz used to measure time in watches and other electronics. Such devices use both properties of piezoelectrics, using electricity to produce a mechanical motion (powering the device) and then using this mechanical motion to produce electricity (generating a signal). The unit of time measured is the natural interval required for electricity to be converted into mechanical energy and back again.

The piezoelectric effect is generally stronger in materials that also exhibit pyroelectricity, and all pyroelectric materials are also piezoelectric. These materials can be used to inter convert between thermal, mechanical, or electrical energy; for instance, after synthesis in a furnace, a pyroelectric crystal allowed to cool under no applied stress generally builds up a static charge of thousands of volts. Such materials are used in motion sensors, where the tiny rise in temperature from a warm body entering the room is enough to produce a measurable voltage in the crystal.

In turn, pyroelectricity is seen most strongly in materials which also display the ferroelectric effect, in which a stable electric dipole can be oriented or reversed by applying an electrostatic field. Pyroelectricity is also a necessary consequence of ferroelectricity. This can be used to store information in ferroelectric capacitors, elements of ferroelectric RAM.

The most common such materials are lead zirconate titanate and barium titanate. Aside from the uses mentioned above, their strong piezoelectric response is exploited in the design of high-frequency loudspeakers, transducers for sonar, and actuators for atomic force and scanning tunneling microscopes.

Positive Thermal Coefficient

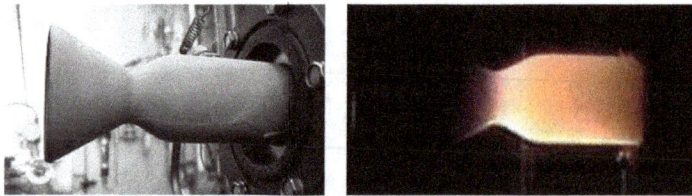

Silicon nitride rocket thruster. Left: Mounted in test stand. Right: Being tested with H_2/O_2 propellants

Increases in temperature can cause grain boundaries to suddenly become insulating in some semiconducting ceramic materials, mostly mixtures of heavy metaltitanates. The critical transition temperature can be adjusted over a wide range by variations in chemistry. In such materials, current will pass through the material until joule heating brings it to the transition temperature, at which point the circuit will be broken and current flow will cease. Such ceramics are used as self-controlled heating elements in, for example, the rear-window defrost circuits of automobiles.

At the transition temperature, the material's dielectric response becomes theoretically infinite. While a lack of temperature control would rule out any practical use of the material near its critical temperature, the dielectric effect remains exceptionally strong even at much higher temperatures. Titanates with critical temperatures far below room temperature have become synonymous with "ceramic" in the context of ceramic capacitors for just this reason.

Optical Properties

Cermax xenon arc lamp with synthetic sapphire output window

Optically transparent materials focus on the response of a material to incoming lightwaves of a range of wavelengths. Frequency selective optical filters can be utilized to alter or enhance the brightness and contrast of a digital image. Guided lightwave transmission via frequency selective waveguides involves the emerging field of fiber optics and the ability of certain glassy compositions as a transmission medium for a range of frequencies simultaneously (multi-mode optical fiber) with little or no interference between competing wavelengths or frequencies. This resonantmode of energy and data transmission via electromagnetic (light) wave propagation, though low powered, is virtually lossless. Optical waveguides are used as components in Integrated optical circuits (e.g. light-emitting diodes, LEDs) or as the transmission medium in local and long haul optical communication systems. Also of value to the emerging materials scientist is the sensitivity of materials to radiation in the thermal infrared (IR) portion of the electromagnetic spectrum. This heat-seeking ability is responsible for such diverse optical phenomena as Night-vision and IR luminescence.

Thus, there is an increasing need in the military sector for high-strength, robust materials which have the capability to transmitlight (electromagnetic waves) in the visible (0.4 – 0.7 micrometers) and mid-infrared (1 – 5 micrometers) regions of the spectrum. These materials are needed for applications requiring transparent armor, including next-generation high-speed missiles and pods, as well as protection against improvised explosive devices (IED).

In the 1960s, scientists at General Electric (GE) discovered that under the right manufacturing conditions, some ceramics, especially aluminium oxide (alumina), could be made translucent. These translucent materials were transparent enough to be used for containing the electrical plasma generated in high-pressuresodium street lamps. During the past two decades, additional types of transparent ceramics have been developed for applications such as nose cones for heat-seekingmissiles, windows for fighter aircraft, and scintillation counters for computed tomography scanners.

In the early 1970s, Thomas Soules pioneered computer modeling of light transmission through translucent ceramic alumina. His model showed that microscopic pores in ceramic, mainly trapped at the junctions of microcrystalline grains, caused light to scatter and prevented true transparency. The volume fraction of these microscopic pores had to be less than 1% for high-quality optical transmission.

This is basically a particle size effect. Opacity results from the incoherent scattering of light at surfaces and interfaces. In addition to pores, most of the interfaces in a typical metal or ceramic object are in the form of grain boundaries which separate tiny regions of crystalline order. When the size of the scattering center (or grain boundary) is reduced below the size of the wavelength of the light being scattered, the scattering no longer occurs to any significant extent.

In the formation of polycrystalline materials (metals and ceramics) the size of the crystalline grains is determined largely by the size of the crystalline particles present in the raw material during formation (or pressing) of the object. Moreover, the size of the grain boundaries scales directly with particle size. Thus a reduction of the original particle size below the wavelength of visible light (~ 0.5 micrometers for shortwave violet) eliminates any light scattering, resulting in a transparent material.

Recently, Japanese scientists have developed techniques to produce ceramic parts that rival the transparency of traditional crystals (grown from a single seed) and exceed the fracture toughness of a single crystal. In particular, scientists at the Japanese firm Konoshima Ltd., a producer of ceramic construction materials and industrial chemicals, have been looking for markets for their transparent ceramics.

Livermore researchers realized that these ceramics might greatly benefit high-powered lasers used in the National Ignition Facility (NIF) Programs Directorate. In particular, a Livermore research team began to acquire advanced transparent ceramics from Konoshima to determine if they could meet the optical requirements needed for Livermore's Solid-State Heat Capacity Laser (SSHCL). Livermore researchers have also been testing applications of these materials for applications such as advanced drivers for laser-driven fusion power plants.

Examples

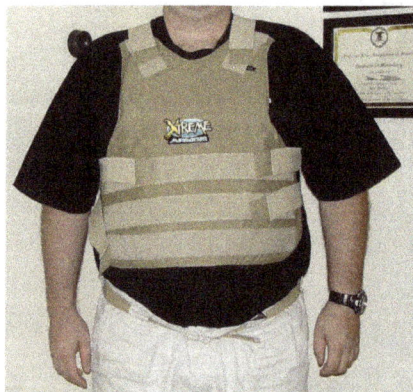

Silicon carbide is used for inner plates of ballistic vests

Porcelain high-voltage insulator

Until the 1950s, the most important ceramic materials were (1) pottery, bricks and tiles, (2) cements and (3) glass. A composite material of ceramic and metal is known as cermet.

Other ceramic materials, generally requiring greater purity in their make-up than those above, include forms of several chemical compounds, including:

- Barium titanate (often mixed with strontium titanate) displays ferroelectricity, meaning that its mechanical, electrical, and thermal responses are coupled to one another and also history-dependent. It is widely used in electromechanicaltransducers, ceramic capacitors, and data storage elements. Grain boundary conditions can create PTC effects in heating elements.

- Bismuth strontium calcium copper oxide, a high-temperature superconductor

- Boron oxide is used in body armor.

- Boron nitride is structurally isoelectronic to carbon and takes on similar physical forms: a graphite-like one used as a lubricant, and a diamond-like one used as an abrasive.

- Earthenware used for domestic ware such as plates and mugs.

- Ferrite is used in the magnetic cores of electrical transformers and magnetic core memory.

- Lead zirconate titanate (PZT) was developed at the United StatesNational Bureau of Standards in 1954. PZT is used as an ultrasonic transducer, as its piezoelectric properties greatly exceed those of Rochelle salt.

- Magnesium diboride (MgB_2) is an unconventional superconductor.

- Porcelain is used for a wide range of household and industrial products.

- Sialon (Silicon Aluminium Oxynitride) has high strength; resistance to thermal shock, chemical and wear resistance, and low density. These ceramics are used in non-ferrous molten metal handling, weld pins and the chemical industry.

- Silicon carbide (SiC) is used as a susceptor in microwave furnaces, a commonly used abrasive, and as a refractory material.

- Silicon nitride (Si_3N_4) is used as an abrasive powder.

- Steatite (magnesium silicates) is used as an electrical insulator.

- Titanium carbide Used in space shuttle re-entry shields and scratchproof watches.

- Uranium oxide (UO_2), used as fuel in nuclear reactors.

- Yttrium barium copper oxide ($YBa_2Cu_3O_{7-x}$), another high temperature super-conductor.

- Zinc oxide (ZnO), which is a semiconductor, and used in the construction of varistors.

- Zirconium dioxide (zirconia), which in pure form undergoes many phase changes between room temperature and practical sintering temperatures, can be chemically "stabilized" in several different forms. Its high oxygen ion conductivity recommends it for use in fuel cells and automotive oxygen sensors. In another variant, metastable structures can impart transformation toughening for mechanical applications; most ceramic knife blades are made of this material.

- Partially stabilised zirconia (PSZ) is much less brittle than other ceramics and is used for metal forming tools, valves and liners, abrasive slurries, kitchen knives and bearings subject to severe abrasion.

Kitchen knife with a ceramic blade

Ceramic Products

By Usage

For convenience, ceramic products are usually divided into four main types; these are shown below with some examples:

- Structural, including bricks, pipes, floor and roof tiles

- Refractories, such as kiln linings, gas fire radiants, steel and glass making crucibles

- Whitewares, including tableware, cookware, wall tiles, pottery products and sanitary ware

- Technical, also known as engineering, advanced, special, and fine ceramics. Such items include:

    o   gas burner nozzles

    o   ballistic protection

    o   nuclear fuel uranium oxide pellets

    o   biomedical implants

    o   coatings of jet engineturbine blades

    o   ceramic disk brake

    o   missile nose cones

    o   bearing (mechanical)

    o   tiles used in the Space Shuttle program

Ceramics Made with Clay

Frequently, the raw materials of modern ceramics do not include clays. Those that do are classified as follows:

- Earthenware, fired at lower temperatures than other types

- Stoneware, vitreous or semi-vitreous

- Porcelain, which contains a high content of kaolin

- Bone china

Classification of Technical Ceramics

Technical ceramics can also be classified into three distinct material categories:

- Oxides: alumina, beryllia, ceria, zirconia

- Nonoxides: carbide, boride, nitride, silicide

- Composite materials: particulate reinforced, fiber reinforced, combinations of oxides and nonoxides.

Each one of these classes can develop unique material properties because ceramics tend to be crystalline.

Applications

- Knife blades: the blade of a ceramic knife will stay sharp for much longer than that of a steel knife, although it is more brittle and can snap from a fall onto a hard surface.

- Carbon-ceramic brake disks for vehicles are resistant to brake fade at high temperatures.

- Advanced composite ceramic and metal matrices have been designed for most modern armoured fighting vehicles because they offer superior penetrating resistance against shaped charges (such as HEAT rounds) and kinetic energy penetrators.

- Ceramics such as alumina and boron carbide have been used in ballistic armored vests to repel large-caliber rifle fire. Such plates are known commonly as small arms protective inserts, or SAPIs. Similar material is used to protect the cockpits of some military airplanes, because of the low weight of the material.

- Ceramics can be used in place of steel for ball bearings. Their higher hardness means they are much less susceptible to wear and typically last for triple the lifetime of a steel part. They also deform less under load, meaning they have less contact with the bearing retainer walls and can roll faster. In very high speed applications, heat from friction during rolling can cause problems for metal bearings, which are reduced by the use of ceramics. Ceramics are also more chemically resistant and can be used in wet environments where steel bearings would rust. In some cases, their electricity-insulating properties may also be valuable in bearings. Two drawbacks to ceramic bearings are a significantly higher cost and susceptibility to damage under shock loads.

- In the early 1980s, Toyota researched production of an adiabaticengine using ceramic components in the hot gas area. The ceramics would have allowed temperatures of over 3000 °F (1650 °C). The expected advantages would have been lighter materials and a smaller cooling system (or no need for one at all), leading to a major weight reduction. The expected increase of fuel efficiency of the engine (caused by the higher temperature, as shown by Carnot's theorem) could not be verified experimentally; it was found that the heat transfer on the hot ceramic cylinder walls was higher than the transfer to a cooler metal wall as the cooler gas film on the metal surface works as a thermal insulator. Thus, despite all of these desirable properties, such engines have not succeeded in production because of costs for the ceramic components and the limited advantages. (Small imperfections in the ceramic material with its low fracture toughness lead to cracks, which can lead to potentially dangerous equipment failure.) Such engines are possible in laboratory settings, but mass production is not feasible with current technology.

- Work is being done in developing ceramic parts for gas turbineengines. Currently, even blades made of advanced metal alloys used in the engines' hot section require cooling and careful limiting of operating temperatures. Turbine engines made with ceramics could operate more efficiently, giving aircraft greater range and payload for a set amount of fuel.

- Recent advances have been made in ceramics which include bioceramics, such as dental implants and synthetic bones. Hydroxyapatite, the natural mineral component of bone, has been made synthetically from a number of biological and chemical sources and can be formed into ceramic materials. Orthopedic implants coated with these materials bond readily to bone and other tissues in the body without rejection or inflammatory reactions so are of great interest for gene delivery and tissue engineering scaffolds. Most hydroxyapatite ceramics are very porous and lack mechanical strength, and are used to coat metal orthopedic devices to aid in forming a bond to bone or as bone fillers. They are also used as fillers for orthopedic plastic screws to aid in reducing the inflammation and increase absorption of these plastic materials. Work is being done to make strong, fully dense nanocrystalline hydroxyapatite ceramic materials for orthopedic weight bearing devices, replacing foreign metal and plastic orthopedic materials with a synthetic, but naturally occurring, bone mineral. Ultimately, these ceramic materials may be used as bone replacements or with the incorporation of protein collagens, synthetic bones.

- High-tech ceramic is used in watchmaking for producing watch cases. The material is valued by watchmakers for its light weight, scratch resistance, durability and smooth touch. IWC is one of the brands that initiated the use of ceramic in watchmaking.

Ceramics in Archaeology

Ceramic artifacts have an important role in archaeology for understanding the culture, technology and behavior of peoples of the past. They are among the most common artifacts to be found at an archaeological site, generally in the form of small fragments of broken pottery called sherds. Processing of collected sherds can be consistent with two main types of analysis: technical and traditional.

Traditional analysis involves sorting ceramic artifacts, sherds and larger fragments into specific types based on style, composition, manufacturing and morphology. By creating these typologies it is possible to distinguish between different cultural styles, the purpose of the ceramic and technological state of the people among other conclusions. In addition, by looking at stylistic changes of ceramics over time is it possible to separate (seriate) the ceramics into distinct diagnostic groups (assemblages). A comparison of ceramic artifacts with known dated assemblages allows for a chronological assignment of these pieces.

The technical approach to ceramic analysis involves a finer examination of the composition of ceramic artifacts and sherds to determine the source of the material and through this the possible manufacturing site. Key criteria are the composition of the clay and the temper used in the manufacture of the article under study: temper is a material added to the clay during the initial production stage, and it is used to aid the subsequent drying process. Types of temper include shell pieces, granite fragments and ground sherd pieces called 'grog'. Temper is usually identified by microscopic examination of the temper material. Clay identification is determined by a process of refiring the ceramic, and assigning a color to it using Munsell Soil Color notation. By estimating both the clay and temper compositions, and locating a region where both are known to occur, an assignment of the material source can be made. From the source assignment of the artifact further investigations can be made into the site of manufacture.

Structure in Materials

Point Lattice

In a point lattice, the following characteristics are obeyed:

- There is a periodic arrangement of points in space. (Figure a)

- In addition, each point must have identical neighbourhood.(Figure b)

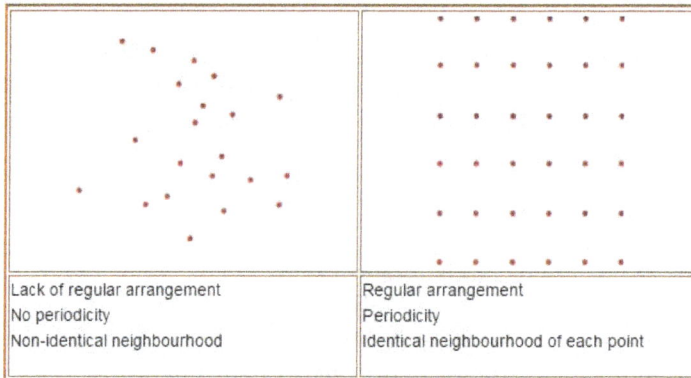

Lack of regular arrangement	Regular arrangement
No periodicity	Periodicity
Non-identical neighbourhood	Identical neighbourhood of each point

(a) Unit-cell representation

Example: Point Lattice

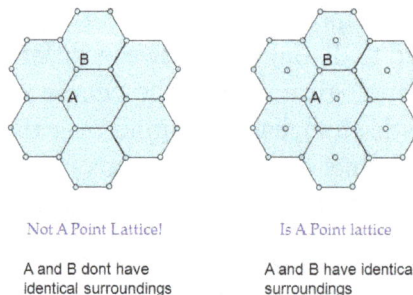

Not A Point Lattice!   Is A Point lattice

A and B dont have identical surroundings   A and B have identical surroundings

(b) Schematics of arrangement of points in space

Unit Cell

- A unit cell is the smallest repeatable unit in a point lattice (Figure c).

- Choice of unit cell shape is not unique.

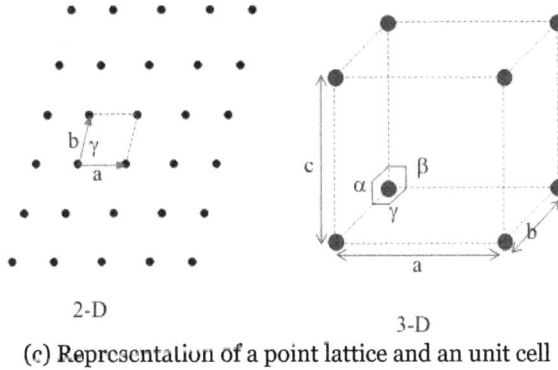

2-D

3-D

(c) Representation of a point lattice and an unit cell

- Unit-cell parameters for a 3-D unit cells

  o   axis lengths: a, b and c

  o   angles:α β and γ

Motif and Crystal Structure

- Crystal structure: a combination of motif and point lattice.

- Motif is defined as a unit or a pattern. For a crystal, it can be an atom, an ion or a group of atoms or ions or a formula unit or formula units. Often it is also called as Basis.

- When motif replaces points in a periodic point lattice, it gives rise to what is called as a crystal with a defined structure.

Motif + Point Lattice = Crystal Structure

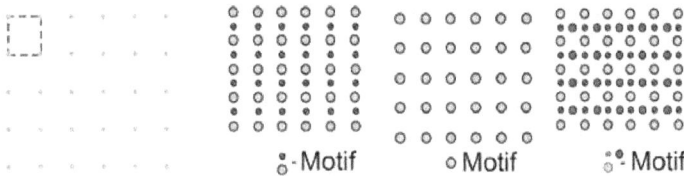

$\frac{\circ}{\circ}$ - Motif   o Motif $\frac{\circ}{\circ}$ Motif

Motif can be defined as a unit of pattern. For a crystal, it is an atom, an ion or a group of atoms or ions or a formula unit or formula units. When motif replaces points in a periodic point lattice, it gives rise to what is called as a crystal with a defined structure.

(d) Formation of a periodic crystal structure

Types of Lattice

Lattice can further be classfied into two types:

- Primitive lattice having one formula unit or one lattice point or one unit of motif per unit cell, and

- Non-primitive lattices having more than one lattice points or more than one unit of motif per unit cell.

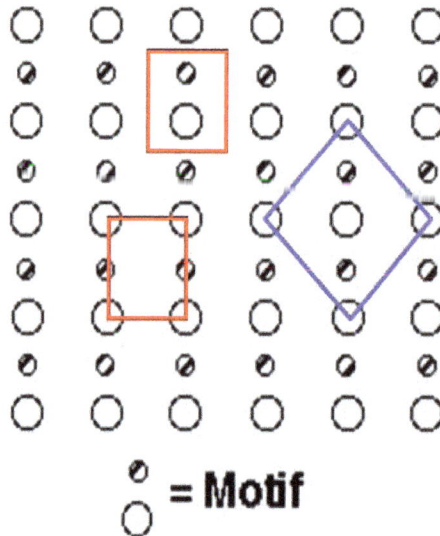

Primitive unit-cell: Consists of one lattice point
Non-primitive cell: More than one lattice point/unit cells
Volume of NP Cell = No of Motifs x Volume of Primitive Unit Cell
Primitive and Non-primitive lattices

Symmetry in Crystals

- Symmetry is an operation which brings the object back to its original confiscation.

- Symmetry elements underlying a point lattice

  o Reflection: reflection across a mirror plane

  o Rotation: rotation around a crystallographic axis by an angle θ such as $360°/\theta$ is an integer of value 1, 2, 3, 4 and 6 and is referred to as n -fold rotation.

  o Inversion: a point at x,y,z becomes its equivalent at $(-x,-y,-z)$

  o Rotation-Inversion: Rotation followed by inversion OR Rotation-Reflection: Rotation followed by reflection

Reflection about a plane: point M, reflects to M,	Rotation: for four fold axis, M, goes to M₂, M₃ and then M₄; for three fold axis M, goes to M₃; two fold axis, M, goes to M₄
Inversion center	Rotation-Inversion center: M, becomes M' due to four fold rotation and then inversion takes M, to M₂

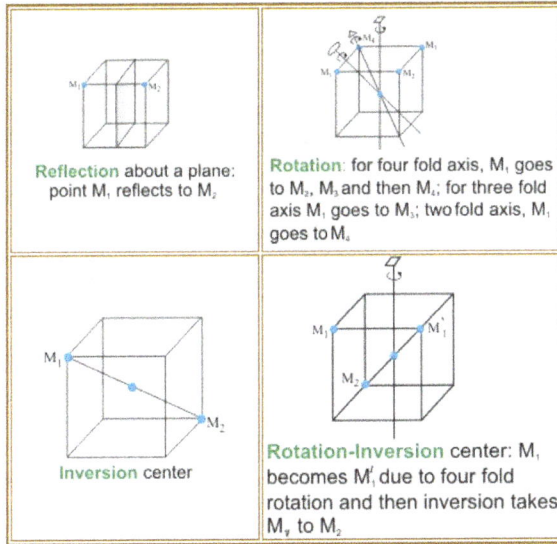

Basic symmetry operations in crystals

Crystal Systems

- As you can see now, the choice of unit cell is not unique and we can define any unit cell of any shape as long it contains one lattice point.

- However, as one starts defining various shapes, we come up with seven categories, called as crystal systems, in which all possible unit cells shapes would fit provided space filling criteria is fulfilled.

- Seven crystal systems are shown below.

	Crystal system and lattice parameters	Minimum symmetry elements
SIMPLE CUBIC (P)	**Cubic** **a=b=c** $\alpha = \beta = y = 90°$	Four 3-fold rotation axes
SIMPLE TETRAGONAL (P)	**Tetragonal** $a = b \neq c$ $\alpha = \beta = y = 90°$	One 4-fold rotation (or rotation-inversion) axis

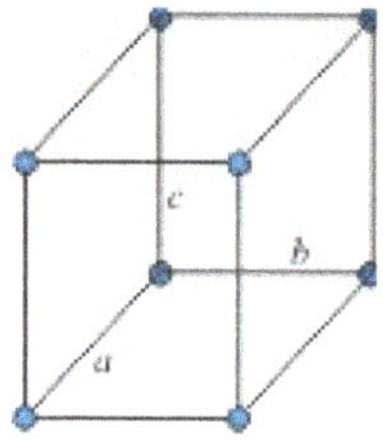 SIMPLE ORTHORHOMBIC (P)	**Orthorhombic** $a \neq b \neq c$ $\alpha = \beta = y = 90°$	Three perpendicular 2-fold rotation (or rotation-inversion) axis
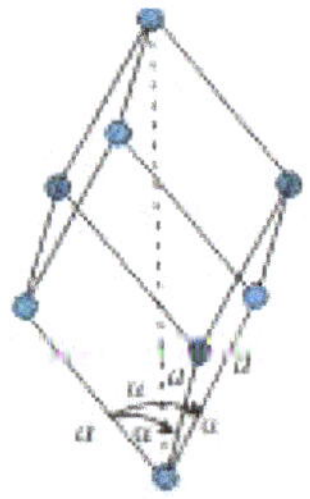 RHOMBOHEDRAL (R)	**Rhombohedral** **A=b=c** $\alpha = \beta = y \neq 90°$	One 3-fold rotation (or rotation-inversion) axis
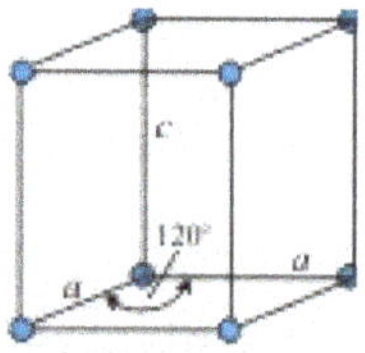 HEXAGONAL (P)	**Hexagonal** $a = b \neq c$ $\alpha = \beta = y \neq 90°$ $y = 120°$	One 6-fold rotation (or rotation-inversion) axis
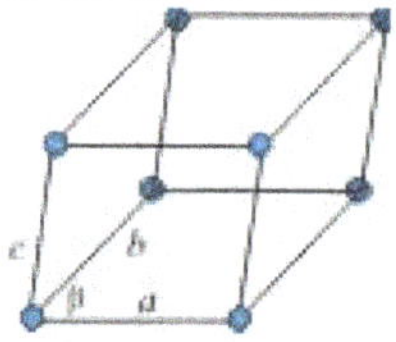 SIMPLE MONOCLICNIC (P)	**Monoclinic** $a \neq b \neq c$ $\alpha = y = 90° \neq \beta$	One 2-fold rotation (or rotation-inversion) axis
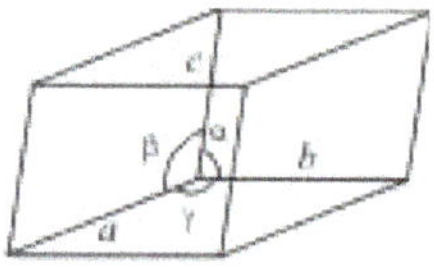 TRICLINIC (P)	**Triclinic** $a \neq b \neq c$ $\alpha \neq y \neq \beta \neq 90°$	None

Seven crystal systems

Bravais Lattices

- Taking seven crystal systems and symmetry elements into account, Bravais came out with the fact that there are a total of 14 Bravais Lattices which are shown below.

Cubic	Tetragonal	Orthorhombic	Rhombohe-dral	Hexagonal	Monoclinic	Triclinic
$a = b = c$ $\alpha = \beta = \gamma = 90$	$a \neq b \neq c$ $\alpha = \beta = \gamma = 90°$	$a \neq b \neq c$ $\alpha = \beta = \gamma = 90°$	$a = b = c$ $\alpha = \beta = \gamma \neq 90°$	$a = b \neq c$ $\alpha = \beta = \gamma \neq 90°$ $\gamma = 120°$	$a \neq b \neq c$ $\alpha = \gamma = 90° \neq \beta$	$a \neq b \neq c$ $\alpha \neq \gamma \neq \beta \neq 90°$

Fourteen Bravais lattices

Planes and Directions

Faces and directions joining atoms in crystals can be best described by Miller Indices (in the names of W. H. Miller) ascribed to determine various planes and directions. While planes are determined little empirically, directions are nothing but vectors.

Crystallographic Planes

- Identification of various faces seen on the crystal

- (h k l) for a plane or {h k l} for identical set of planes

- A crystallographic plane in a crystal satisfies the following equation

$$\frac{h}{a}x + \frac{k}{b}x + \frac{l}{c}z = 1 \qquad (1)$$

- h/a, k/b, and c/l are the intercepts of the plane on x, y, and z axes.

- a, b, c are the unit cell lengths

- h, k, l are integers called Miller indices and the plane is represented as (h, k, l)

- Any negative indices in Miller indices of a plane is written with a bar on top such as h.

Directions

- These are basically atomic directions in the crystal.

- Miller indices are [u , v , w] for a direction or < u , v , w > for identical set of directions where u , v , w are integers

- Vector components of the direction resolved along each of the crystal axis reduced to smallest set of integers

How to Find Miller Indices

1) Find the intercepts of the plane with the crystal axes. Express them, as integral multiples of the basis vectors.

2) Take the reciprocals of the three integers found in step 1. If possible reduce these to smallest set of integers *h, k and l*.

3) Label the plane (*hkl*)

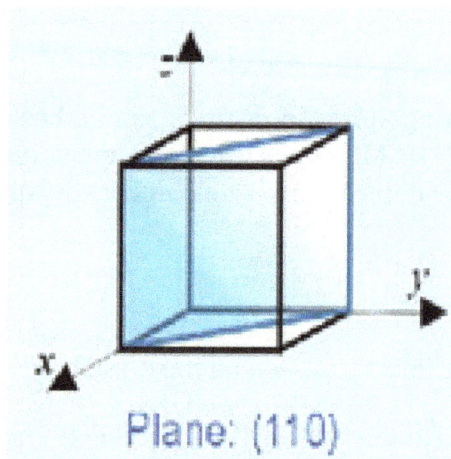

Plane: (110)

Interplanar angle is given by (cubic only)

$$\cos \theta = \frac{h_1 h_2 + k_1 k_2 + l_1 l_2}{\sqrt{h_1^2 + k_1^2 + l_1^2} \sqrt{h_2^2 + k_2^2 + l_2^2}}$$

Interplanar space is given by (cubic)

$$d_{hkl} = \frac{a}{\sqrt{h^2 + k^2 + l^2}}$$

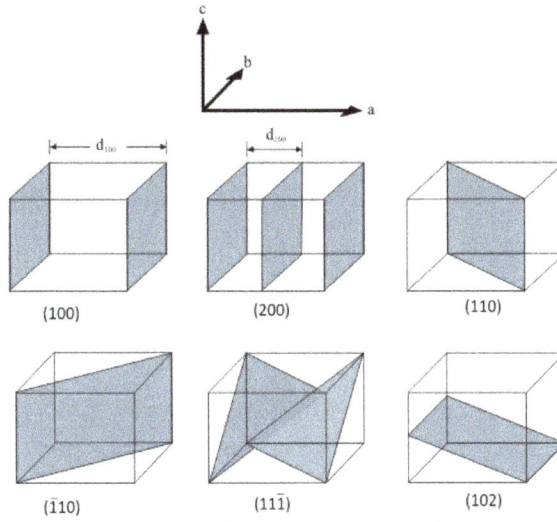

Planes and Directions in Crystals

Crystal Directions

Planes and Directions in Crystals

Bonding in Materials

Primary Bonding

There are three types of primary bonding mechanisms: metallic, covalent and ionic bonding.

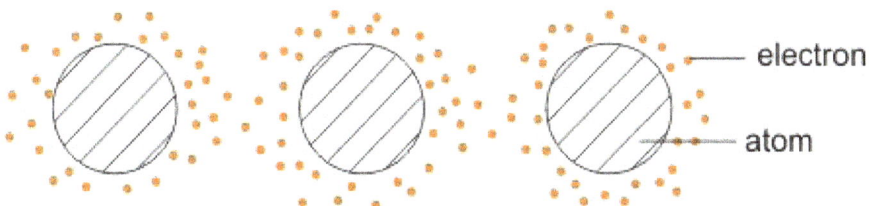

Metallic Bonding

Metallic bonding :

- This kind of bonding is characterized by presence of a sea of electrons around atoms in metal giving rise to flexible bonds, good malleability, high electrical and thermal conductivity. Most metals such as Ni, Fe, Cu, Au, Ag etc exhibit this kind of bonding.

Covalent Bonding:

- In this bonding, atoms share their outer shell unpaired electrons leading to a stronger and directional bonding.

- Examples of materials showing this bonding are mainly group IV elements and compounds such as Si, C, Ge, and SiC and gases like methane.

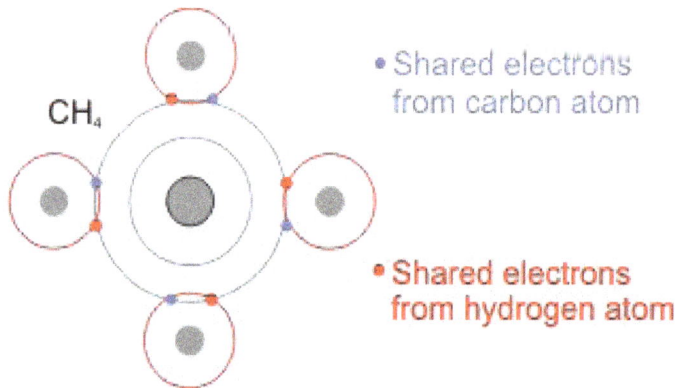

Schematic of covalent bonding

Ionic Bonding :

- This bonding occurs due to large differences in the electronegativities of two elements, for example in NaCl, MgO etc.

- This type of bonding typically leads to high bond energies, high bond strength, high modulus, brittle nature, generally low thermal and electrical conductivities making them excellent insulators.

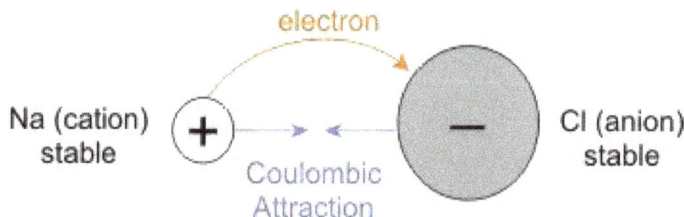

(j) Schematic of ionic bonding

Secondary Bonding :

- It arises from the interaction between charge dipoles.

Fluctuating Dipoles:

- Observed in gases like hydrogen.

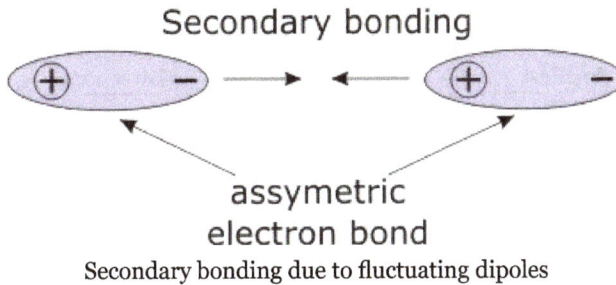

Secondary bonding due to fluctuating dipoles

Permanent Dipole Moment Induced

- Induced to permanent dipoles in the materials

- General case

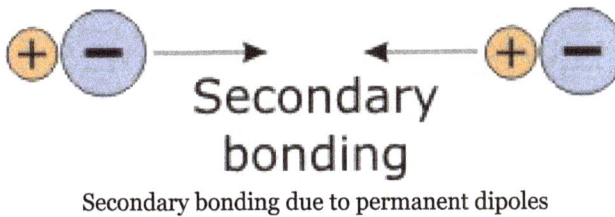

Secondary bonding due to permanent dipoles

Examples are materials like polymers.

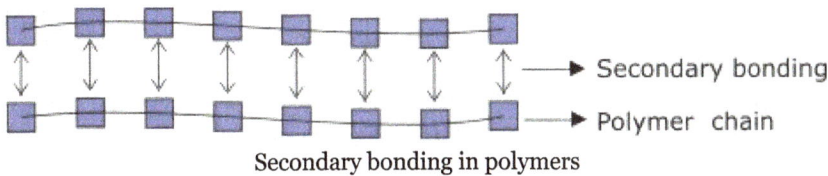

Secondary bonding in polymers

Bonding, Bond Energy and General Remarks

Type	Bond Energy	Comments
Ionic	Large magnitude (Large T_m . large E and small a) Example: MgO – 1000 kJ / mol. T_m - 2800°C	Non-directional (Typically Ceramics)
Covalent	Variable for materials such as Si,Ge Large for Diamond (Carbon) and small for Bismuth Typically Large T_m and LargeE Example: SI – 450 kJ/mol; T_m - 1410° C	Directional (Typically Semiconductors, Polymers and some Ceramics)

Metallic	Variable Large: Tungsten (W) Small: Mercury (Hg) Moderate: Al: 68 kJ/mol, $T_m \sim 670°$ C	Non-directional (Metals)
Secondary	Smallest Characterised by low T_m . low E and large α	Directional Inter-chain (Polymer) Inter-molecular

Symbols: T_m: Melting point, a: Coefficient of thermal expansion, E: Elastic modulus

Cubic Crystal System

A rock containing three crystals of pyrite (FeS_2). The crystal structure of pyrite is primitive cubic, and this is reflected in the cubic symmetry of its natural crystal facets.

In crystallography, the cubic (or isometric) crystal system is a crystal system where the unit cell is in the shape of a cube. This is one of the most common and simplest shapes found in crystals and minerals.

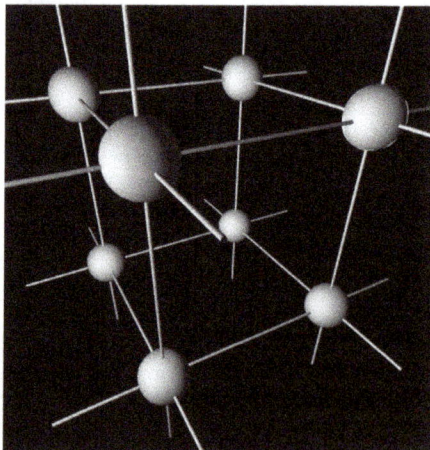

A network model of a primitive cubic system

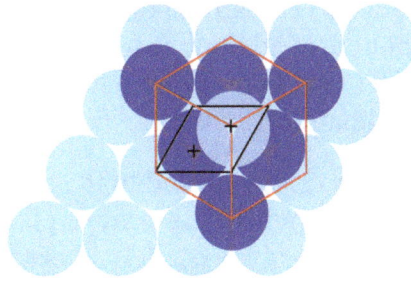

The primitive and cubic close-packed (also known as face-centered cubic) unit cells

There are three main varieties of these crystals:

- Primitive cubic (abbreviated *cP* and alternatively called *simple cubic*)

- Body-centered cubic (abbreviated *cI* or *bcc*),

- Face-centered cubic (abbreviated *cF* or *fcc*, and alternatively called *cubic close-packed* or *ccp*)

Each is subdivided into other variants listed below. Note that although the *unit cell* in these crystals is conventionally taken to be a cube, the *primitive* unit cell often is not. This is related to the fact that in most cubic crystal systems, there is more than one atom per cubic unit cell.

A classic isometric crystal has square or pentagonal faces.

Bravais Lattices

The three Bravais lattices in the cubic crystal system are:

Bravais lattice	Primitive cubic	Body-centered cubic	Face-centered cubic
Pearson symbol	cP	cI	cF
Unit cell			

The primitive cubic system (cP) consists of one lattice point on each corner of the cube. Each atom at a lattice point is then shared equally between eight adjacent cubes, and the unit cell therefore contains in total one atom ($\frac{1}{8} \times 8$).

The body-centered cubic system (cI) has one lattice point in the center of the unit cell in addition to the eight corner points. It has a net total of 2 lattice points per unit cell ($\frac{1}{8} \times 8 + 1$).

The face-centered cubic system (cF) has lattice points on the faces of the cube, that each gives exactly one half contribution, in addition to the corner lattice points, giving a total of 4 lattice points per unit cell ($\frac{1}{8}\times 8$ from the corners plus $\frac{1}{2}\times 6$ from the faces). Each sphere in a cF lattice has coordination number 12.

The face-centered cubic system is closely related to the hexagonal close packed (HCP) system, and the two systems differ only in the relative placements of their hexagonal layers. The plane of a face-centered cubic system is a hexagonal grid.

Attempting to create a C-centered cubic crystal system (i.e., putting an extra lattice point in the center of each horizontal face) would result in a simple tetragonal Bravais lattice.

Crystal Classes

The *isometric crystal system class* names, examples, Schönflies notation, Hermann–Mauguin notation, point groups, International Tables for Crystallography space group number, orbifold, type, and space groups are listed in the table below. There are a total 36 cubic space groups.

#	Point group					Type Example	Space groups		
	Class	Schön.	Intl	Orb.	Cox.		Primitive	Face-centered	Body-centered
195–197	Tetartoidal	T	23	332	$[3,3]^+$	enantiomorphic (Ullmannite)	P23	F23	I23
198–199							$P2_13$		$I2_13$
200–205	Diploidal	T_h	$2/m\bar{3}$ ($m\bar{3}$)	3*2	$[3^+,4]$	centrosymmetric (Pyrite)	$Pm\bar{3}$, $Pn\bar{3}$	$Fm\bar{3}$, $Fd\bar{3}$	$I\bar{3}$
205–206							$Pa\bar{3}$		$Ia\bar{3}$
207–211	Gyroidal	O	432	432	$[3,4]^+$	enantiomorphic (Petzite)	P432, $P4_232$	F432, $F4_132$	I432
212–214							$P4_332$, $P4_132$		$I4_132$
215–217	Hextetrahedral	T_d	$\bar{4}3m$	*332	$[3,3]$	(Sphalerite)	$P\bar{4}3m$	$F\bar{4}3m$	$I\bar{4}3m$
218–220							$P\bar{4}3n$	$F\bar{4}3c$	$I\bar{4}3d$
221–230	Hexoctahedral	O_h	$4/m\bar{3}$ $2/m$ ($m\bar{3}m$)	*432	$[3,4]$	centrosymmetric (Galena)	$Pm\bar{3}m$, $Pn\bar{3}n$, $Pm\bar{3}n$, $Pn\bar{3}m$	$Fm\bar{3}m$, $Fm\bar{3}c$, $Fd\bar{3}m$, $Fd\bar{3}c$	$Im\bar{3}m$, $Ia\bar{3}d$

Other terms for hexoctahedral are: normal class, holohedral, ditesseral central class, galena type.

Voids in the Unit Cell

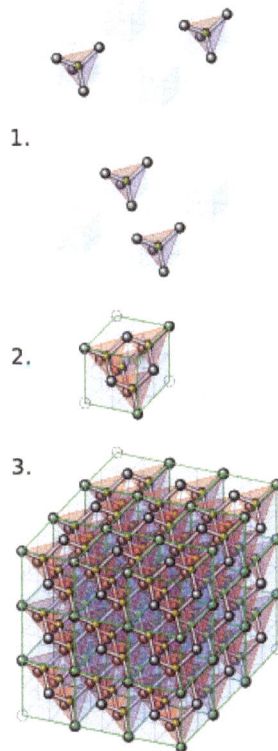

Visualisation of a diamond cubic unit cell: 1. Components of a unit cell, 2. One unit cell, 3. A lattice of 3 x 3 x 3 unit cells

A simple cubic unit cell has a single cubic void in the center.

A body-centered cubic unit cell has six octahedral voids located at the center of each face of the unit cell, and twelve further ones located at the midpoint of each edge of the same cell, for a total of six net octahedral voids. Additionally, there are 24 tetrahedral voids located in a square spacing around each octahedral void, for a total of twelve net tetrahedral voids. These tetrahedral voids are not local maxima and are not technically voids, but they do occasionally appear in multi-atom unit cells.

A face-centered cubic unit cell has eight tetrahedral voids located midway between each corner and the center of the unit cell, for a total of eight net tetrahedral voids. Additionally, there are twelve octahedral voids located at the midpoints of the edges of the unit cell as well as one octahedral hole in the very center of the cell, for a total of four net octahedral voids.

One important characteristic of a crystalline structure is its atomic packing factor. This is calculated by assuming that all the atoms are identical spheres, with a radius large enough that each sphere abuts on the next. The atomic packing factor is the proportion of space filled by these spheres.

Assuming one atom per lattice point, in a primitive cubic lattice with cube side length a, the sphere radius would be $\frac{a}{2}$ and the atomic packing factor turns out to be about 0.524 (which is quite low). Similarly, in a bcc lattice, the atomic packing factor is 0.680, and in fcc it is 0.740. The fcc value is the highest theoretically possible value for any lattice, although there are other lattices which also achieve the same value, such as hexagonal close packed and one version of tetrahedral bcc.

As a rule, since atoms in a solid attract each other, the more tightly packed arrangements of atoms tend to be more common. (Loosely packed arrangements do occur, though, for example if the orbital hybridization demands certain bond angles.) Accordingly, the primitive-cubic structure, with especially low atomic packing factor, is rare in nature, but is found in polonium. The bcc and fcc, with their higher densities, are both quite common in nature. Examples of bcc include iron, chromium, tungsten, and niobium. Examples of fcc include aluminium, copper, gold and silver.

Multi-element Compounds

Compounds that consist of more than one element (e.g. binary compounds) often have crystal structures based on a cubic crystal system. Some of the more common ones are listed here.

Caesium Chloride Structure

The space group of the caesium chloride (CsCl) structure is called Pm$\bar{3}$m (in Hermann–Mauguin notation), or "221" (in the International Tables for Crystallography). The Strukturbericht designation is "B2".

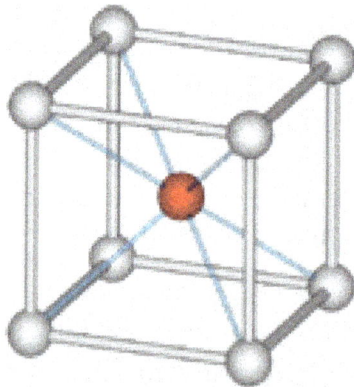

A caesium chloride unit cell. The two colors of spheres represent the two types of atoms.

One structure is the "interpenetrating primitive cubic" structure, also called the "caesium chloride" structure. Each of the two atom types forms a separate primitive cubic lattice, with an atom of one type at the center of each cube of the other type. Altogether, the arrangement of atoms is the same as body-centered cubic, but with alternating types of atoms at the different lattice sites. Alternately, one could view this lattice as a simple cubic structure with a secondary atom in its cubic void.

In addition to caesium chloride itself, the structure also appears in certain other alkali halides when prepared at low temperatures or high pressures. Generally, this structure is more likely to be formed from two elements whose ions are of roughly the same size (for example, ionic radius of Cs^+ = 181 pm, and Cl^- = 167 pm).

The coordination number of each atom in the structure is 8: the central cation is coordinated to 8 anions on the corners of a cube as shown, and similarly, the central anion is coordinated to 8 cations on the corners of a cube.

Other compounds showing caesium chloride like structure are CsBr, CsI, high-temp RbCl, AlCo, AgZn, BeCu, MgCe, RuAl and SrTl.

Rock-salt Structure

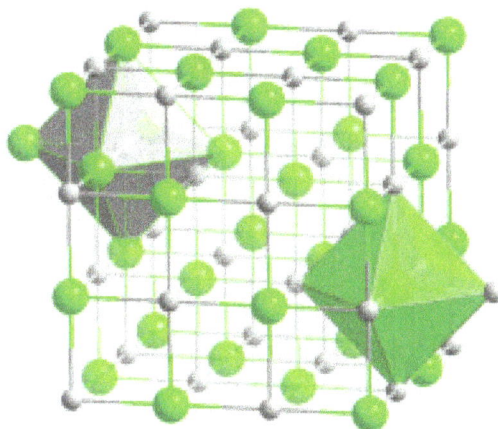

The rock-salt crystal structure. Each atom has six nearest neighbors, with octahedral geometry.

The space group of the rock-salt (NaCl) structure is called $Fm\bar{3}m$ (in Hermann–Mauguin notation), or "225" (in the International Tables for Crystallography). The Strukturbericht designation is "B1".

In the rock-salt or sodium chloride (halite) structure, each of the two atom types forms a separate face-centered cubic lattice, with the two lattices interpenetrating so as to form a 3D checkerboard pattern. Alternately, one could view this structure as a face-centered cubic structure with secondary atoms in its octahedral holes.

Examples of compounds with this structure include sodium chloride itself, along with almost all other alkali halides, and "many divalent metal oxides, sulfides, selenides, and tellurides". More generally, this structure is more likely to be formed if the cation is somewhat smaller than the anion (a cation/anion radius ratio of 0.414 to 0.732).

The coordination number of each atom in this structure is 6: each cation is coordinated to 6 anions at the vertices of an octahedron, and similarly, each anion is coordinated to 6 cations at the vertices of an octahedron.

The interatomic distance (distance between cation and anion, or half the unit cell length a) in some rock-salt-structure crystals are: 2.3 Å (2.3×10^{-10} m) for NaF, 2.8 Å for NaCl, and 3.2 Å for SnTe.

Other compounds showing rock salt like structure are LiF, LiCl, LiBr, LiI, NaF, NaBr, NaI, KF, KCl, KBr, KI, RbF, RbCl, RbBr, RbI, CsF, MgO, PbS, AgF, AgCl, AgBr and ScN.

Zincblende Structure

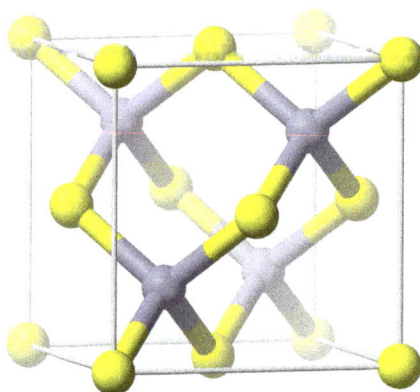

A zincblende unit cell

The space group of the Zincblende structure is called $F\bar{4}3m$ (in Hermann–Mauguin notation), or 216. The Strukturbericht designation is "B3".

The Zincblende structure (also written "zinc blende") is named after the mineral zinc-blende (sphalerite), one form of zinc sulfide (β-ZnS). As in the rock-salt structure, the two atom types form two interpenetrating face-centered cubic lattices. However, it differs from rock-salt structure in how the two lattices are positioned relative to one another. The zincblende structure has tetrahedralcoordination: Each atom's nearest neighbors consist of four atoms of the opposite type, positioned like the four vertices of a regular tetrahedron. Altogether, the arrangement of atoms in zincblende structure is the same as diamond cubic structure, but with alternating types of atoms at the different lattice sites.

Examples of compounds with this structure include zincblende itself, lead(II) nitrate, many compound semiconductors (such as gallium arsenide and cadmium telluride), and a wide array of other binary compounds.

Other compounds showing zinc blende-like structure are α-AgI, β-BN, diamond, CuBr, β-CdS, BP and BAs.

Weaire–Phelan Structure

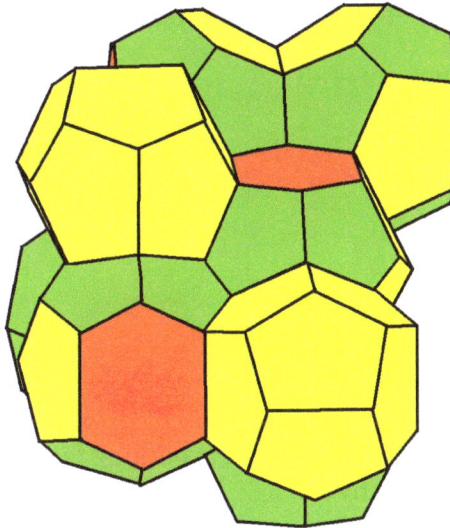

Weaire–Phelan structure

The Weaire–Phelan structure has Pm3n (223) symmetry.

It has 3 orientations of stacked tetradecahedrons with pyritohedral cells in the gaps. It is found as a crystal structure in chemistry where it is usually known as the "Type I clathrate structure". Gas hydrates formed by methane, propane, and carbon dioxide at low temperatures have a structure in which water molecules lie at the nodes of the Weaire–Phelan structure and are hydrogen bonded together, and the larger gas molecules are trapped in the polyhedral cages.

Simple Cubic Structure

- Simplest structure crystallographically but in the entire periodic table only polonium (Po) possesses this structure.

- Structure contains only one atom per unit-cell.

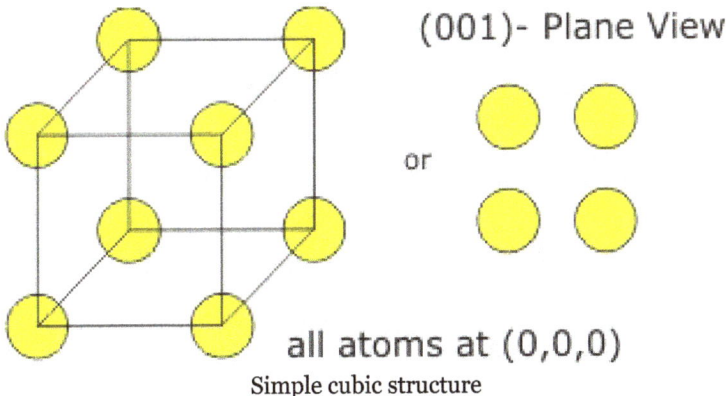

(001)- Plane View

or

all atoms at (0,0,0)

Simple cubic structure

Body Centered Cubic or BCC Structure

- Many metals like W, Fe (room temperature form) possess BCC structure.

- Contains 2 atoms per unit-cell

BCC Structure

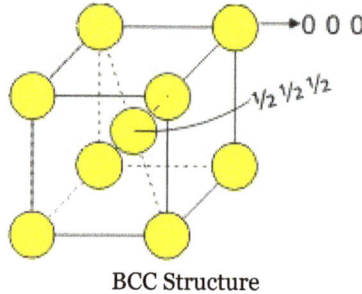

One of the important parameters of interest is packing factor, determining how loosely or densely a structure is packed by atoms.

Packing Factor: Volume of all atoms in one unit cell divided by Volume of one unit-cell

If r is the atomic radii in these structures, then

$$\text{Packing Factor (Simple Cubic)} = \frac{1 \times \frac{4}{3} \times \pi r^3}{(2r)^3} = 0.52$$

$$\text{Packing Factor (BCC)} = \frac{2 \times \frac{4}{3} \times \pi r^3}{\left(\frac{4}{\sqrt{3}} r\right)^3} = 0.68$$

Closed Packed Structures

- Each atom has 12 nearest neighbours touching the atom to each other.

  ABC ABC ABC . . . stacking leads to hexagonal closed packed (HCP) structure. The A or B planes are closed packed c-plane or (001) planes of hexagonal structure.

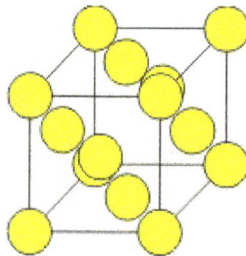

FCC/CCP Structure   HCP Structure

FCC and HCP Structures

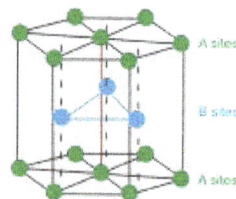

Now you can work out yourself that packing factor of both FCC and HCP is 0.74.

Interstices in Structures

- Since the unit cell is not completely packed as packing efficiency in the previous structures is less than 100%, there are empty spaces inside which are called as interstices.

- These interstices are very useful because there can contain smaller atoms which modify the properties of materials tremendously, such as Carbon (C) in Iron (Fe) makes steel and makes iron stronger.

Interstices in FCC Structure

- Tetrahedral Interstices

    o   2 per atom

- Octahedral Interstices

    o   1 per atom

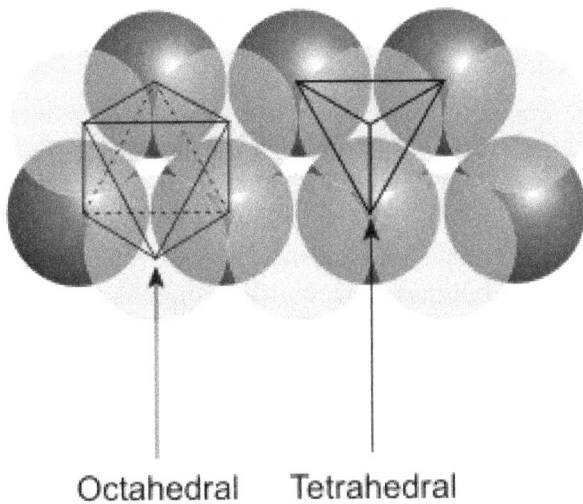

Tetrahedral Interstices

Four-fold coordination

$\frac{1}{4},\frac{1}{4},\frac{1}{4}$ sites in FCC

Octahedral Interstices

Six-fold coordination

$\frac{1}{2},\frac{1}{2},\frac{1}{2}$ and $\frac{1}{2},0,0$ sites in FCC

Octahedral   Tetrahedral

Interstices in a FCC structure

So, by simple geometry, you can also estimate the size of the largest interstitial atom that would fit in these interstices without distorting them.

$$r_{tet} = 0.225 * r$$

$$r_{tet} = 0.225 * r$$

Interstices in BCC Structures

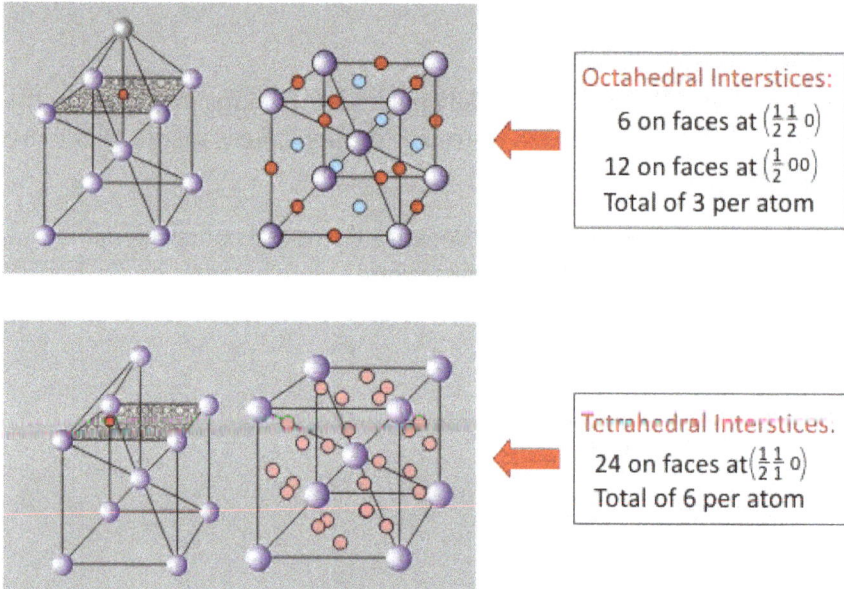

Octahedral Interstices:
6 on faces at $\left(\frac{1}{2}\frac{1}{2}\,0\right)$
12 on faces at $\left(\frac{1}{2}\,00\right)$
Total of 3 per atom

Tetrahedral Interstices:
24 on faces at $\left(\frac{1}{2}\frac{1}{2}\,0\right)$
Total of 6 per atom

Interstices in a BCC structure

Structure of Covalent Ceramics

Most ceramic materials are neither purely covalently or ionically bonded materials. In most ionically bonded materials, there is a significant level of covalency which is decreases as the difference between the electronegativities of cations and anions increases. While covalent bonding is prevalent among the group IV solids such as diamond and many other compound semiconductors, most ceramics such as NaCl, MgO, $BaTiO_3$, Fe_3O_4 etc are predominantly ionically bonded. Covalent bonding, as we saw, arises from the sharing of orbitals and as a result materials with this type of bonding are characterized by significant hybridization of orbitals and directionality of the bonds which play a crucial role in determining the crystal structure. In contrast, ionically bonded solids are predominantly based on the size difference between the cations and the anions and the formation of structures in them is determined by a set of rules called as Pauling's Rules.

In this section, we will understand the structures of a few covalently bonded materials with emphasis on the Diamond structure.

Diamond Cubic

The diamond cubic crystal structure is a repeating pattern of 8 atoms that certain materials may adopt as they solidify. While the first known example was diamond, other elements in group 14 also adopt this structure, including α-tin, the semiconductors silicon and germanium, and silicon/germanium alloys in any proportion.

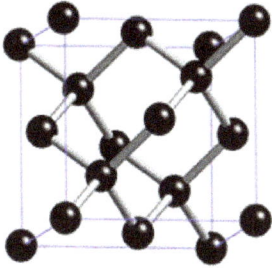

Unit cell of the diamond cubic crystal structure

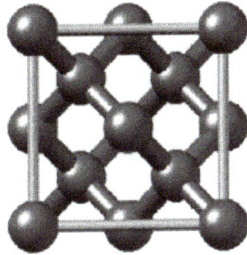

Model of the diamond cubic crystal structure

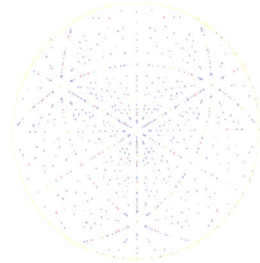

Pole figure in stereographic projection of the diamond lattice showing the 3-fold symmetry along the direction.

Crystallographic Structure

Diamond cubic is in the Fd3̄m space group, which follows the face-centered cubicBravais lattice. The lattice describes the repeat pattern; for diamond cubic crystals this lattice is "decorated" with a *motif* of two tetrahedrallybonded atoms in each primitive cell, separated by 1/4 of the width of the unit cell in each dimension. The diamond lattice can be viewed as a pair of intersecting face-centered cubic lattices, with each separated by 1/4 of the width of the unit cell in each dimension. Many compound semiconductors such as gallium arsenide, β-silicon carbide, and indium antimonide adopt the analogous zincblende structure, where each atom has nearest neighbors of an unlike element. Zincblende's space group is F4̄3m, but many of its structural properties are quite similar to the diamond structure.

The atomic packing factor of the diamond cubic structure (the proportion of space that would be filled by spheres that are centered on the vertices of the structure and are as large as possible without overlapping) is $\pi\sqrt{3}/16 \approx 0.34$, significantly smaller (indicating a less dense structure) than the packing factors for the face-centered and body-centered cubic lattices. Zincblende structures have higher packing factors than 0.34 depending on the relative sizes of their two component atoms.

The first-, second-, third-, fourth- and fifth-nearest-neighbor distances in units of the cubic lattice constant are $\sqrt{3}/4$, $\sqrt{2}/2$, $\sqrt{11}/4$, 1 and $\sqrt{19}/4$, respectively.

Mathematical Structure

Mathematically, the points of the diamond cubic structure can be given coordinates as a subset of a three-dimensional integer lattice by using a cubic unit cell four units across. With these coordinates, the points of the structure have coordinates (x, y, z) satisfying the equations:

$$x = y = z \pmod 2, \text{ and}$$

$$x + y + z = 0 \text{ or } 1 \pmod 4.$$

There are eight points (modulo 4) that satisfy these conditions:

$$(0,0,0), (0,2,2), (2,0,2), (2,2,0),$$

$$(3,3,3), (3,1,1), (1,3,1), (1,1,3)$$

All of the other points in the structure may be obtained by adding multiples of four to the x, y, and z coordinates of these eight points. Adjacent points in this structure are at distance $\sqrt{3}$ apart in the integer lattice; the edges of the diamond structure lie along the body diagonals of the integer grid cubes. This structure may be scaled to a cubical unit cell that is some number a of units across by multiplying all coordinates by $a/4$.

Alternatively, each point of the diamond cubic structure may be given by four-dimensional integer coordinates whose sum is either zero or one. Two points are adjacent in the diamond structure if and only if their four-dimensional coordinates differ by one in a single coordinate. The total difference in coordinate values between any two points (their four-dimensional Manhattan distance) gives the number of edges in the shortest path between them in the diamond structure. The four nearest neighbors of each point may be obtained, in this coordinate system, by adding one to each of the four coordinates, or by subtracting one from each of the four coordinates, accordingly as the coordinate sum is zero or one. These four-dimensional coordinates may be transformed into three-dimensional coordinates by the formula

$$(a, b, c, d) \rightarrow (a + b - c - d, a - b + c - d, -a + b + c - d).$$

Because the diamond structure forms a distance-preserving subset of the four-dimensional integer lattice, it is a partial cube.

Yet another coordinatization of the diamond cubic involves the removal of some of the edges from a three-dimensional grid graph. In this coordinatization, which has a distorted geometry from the standard diamond cubic structure but has the same topological structure, the vertices of the diamond cubic are represented by all possible 3d grid points and the edges of the diamond cubic are represented by a subset of the 3d grid edges.

The diamond cubic is sometimes called the "diamond lattice" but it is not, mathematically, a lattice: there is no translational symmetry that takes the point $(0,0,0)$ into the point $(3,3,3)$, for instance. However, it is still a highly symmetric structure: any incident pair of a vertex and edge can be transformed into any other incident pair by a congruence of Euclidean space. Moreover the diamond crystal as a network in space has a strong isotropic property. Namely, for any two vertices x and y of the crystal net, and for any ordering of the edges adjacent to x and any ordering of the edges adjacent to y, there is a net-preserving congruence taking x to y and each x-edge to the similarly ordered y-edge. Another (hypothetical) crystal with this property is the Laves graph (also called the K_4 crystal, (10,3)-a, or the diamond twin).

Mechanical Properties

The compressive strength and hardness of diamond and various other materials, such as boron nitride, is attributed to the diamond cubic structure.

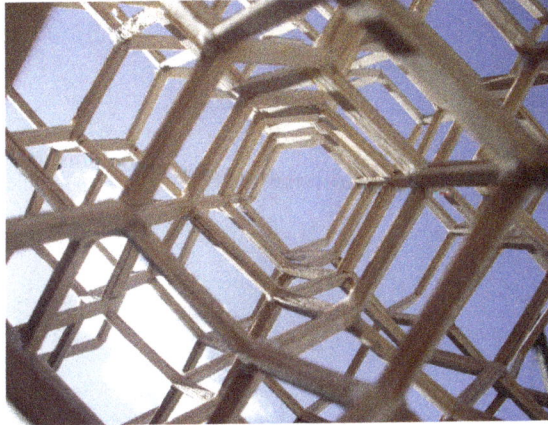

Example of a diamond cubic truss system for resisting compression

Similarly truss systems that follow the diamond cubic geometry have a high capacity to withstand compression, by minimizing the unbraced length of individual struts. The diamond cubic geometry has also been considered for the purpose of providing structural rigidity though structures composed of skeletal triangles, such as the octet truss, have been found to be more effective for this purpose.

Graphite

Graphite, archaically referred to as plumbago, is a crystallineallotrope of carbon, a semimetal and a native element mineral. Graphite is the most stable form of carbon under standard conditions. Therefore, it is used in thermochemistry as the standard state for defining the heat of formation of carbon compounds.

Types and Varieties

The principal types of natural graphite, each occurring in different types of ore deposits are:

- Crystalline flake graphite (or flake graphite) occurs as isolated, flat, plate-like particles with hexagonal edges if unbroken. When broken the edges can be irregular or angular;

- Amorphous graphite: very fine flake graphite is sometimes called amorphous;

- Lump graphite (or vein graphite) occurs in fissure veins or fractures and appears as massive platy intergrowths of fibrous or acicular crystalline aggregates, and is probably hydrothermal in origin.

- Highly ordered pyrolytic graphite refers to graphite with an angular spread between the graphite sheets of less than 1°.

- The name "graphite fiber" is sometimes used to refer to carbon fibers or carbon fiber-reinforced polymer.

Occurrence

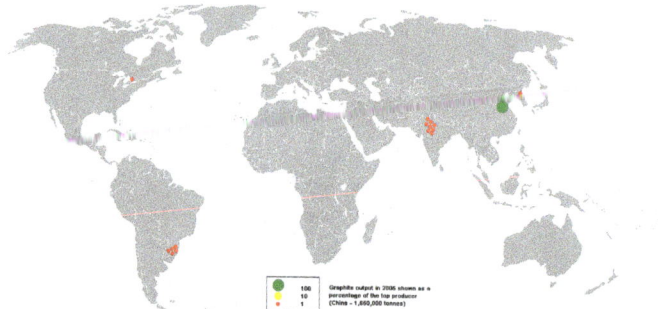

Graphite output in 2005

Graphite occurs in metamorphic rocks as a result of the reduction of sedimentary carbon compounds during metamorphism. It also occurs in igneous rocks and in meteorites. Minerals associated with graphite include quartz, calcite, micas and tourmaline. In meteorites it occurs with troilite and silicate minerals. Small graphitic crystals in meteoritic iron are called cliftonite.

According to the United States Geological Survey (USGS), world production of natural graphite in 2012 was 1,100,000 tonnes, of which the following major exporters are: China (750 kt), India (150 kt), Brazil (75 kt), North Korea (30 kt) and Canada (26 kt). Graphite is not mined in the United States, but U.S. production of synthetic graphite in 2010 was 134 kt valued at $1.07 billion.

Properties

Structure

Graphite has a layered, planar structure. The individual layers are called graphene. In each layer, the carbon atoms are arranged in a honeycomb lattice with separation of 0.142 nm, and the distance between planes is 0.335 nm. Atoms in the plane are bonded covalently, with only three of the four potential bonding sites satisfied. The fourth electron is free to migrate in the plane, making graphite electrically conductive. However, it does not conduct in a direction at right angles to the plane. Bonding between layers is via weak van der Waals bonds, which allows layers of graphite to be easily separated, or to slide past each other.

The two known forms of graphite, *alpha* (hexagonal) and *beta* (rhombohedral), have very similar physical properties, except for that the graphene layers stack slightly differently. The alpha graphite may be either flat or buckled. The alpha form can be converted to the beta form through mechanical treatment and the beta form reverts to the alpha form when it is heated above 1300 °C.

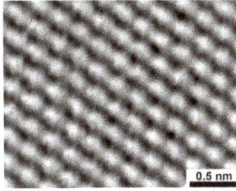

Scanning tunneling microscope image of graphite surface atoms

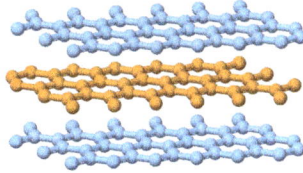

Side view of layer stacking

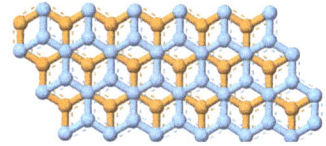

Plane view of layer stacking

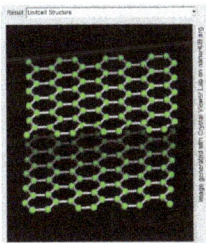

Ball-and-stick model of graphite (two graphene layers)

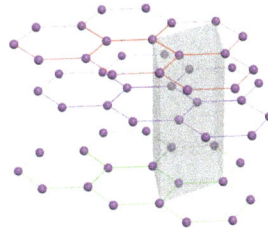

View of the unit cell in three layers of graphene (note that this is a slightly different unit cell from the one to the left)

Graphite's unit cell

Other Properties

Graphite plates and sheets, 10–15 cm high, Mineral specimen from Kimmirut, Baffin Island.

The acoustic and thermal properties of graphite are highly anisotropic, since phonons propagate quickly along the tightly-bound planes, but are slower to travel from one plane to another. Graphite's high thermal stability and electrical and thermal conductivity facilitate its widespread use as electrodes and refractories in high temperature material processing applications. However, in oxygen containing atmospheres graphite readily oxidizes to form CO_2 at temperatures of 700 °C and above.

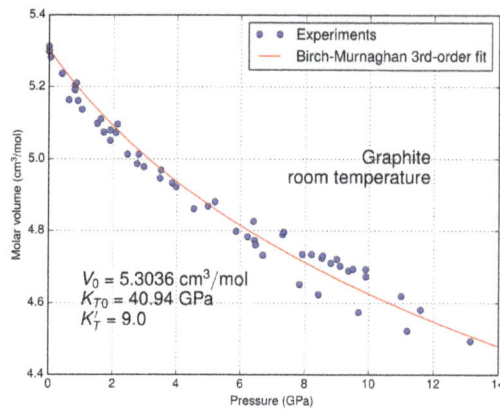

Molar volume vs. pressure at room temperature.

Graphite is an electric conductor, consequently, useful in such applications as arc lampelectrodes. It can conduct electricity due to the vast electrondelocalization within the carbon layers (a phenomenon called aromaticity). These valence electrons are free to move, so are able to conduct electricity. However, the electricity is primarily conducted within the plane of the layers. The conductive properties of powdered graphite allows its use as pressure sensor in carbon microphones.

Graphite and graphite powder are valued in industrial applications for their self-lubricating and dry lubricating properties. There is a common belief that graphite's lubricating properties are solely due to the loose interlamellar coupling between sheets in the structure. However, it has been shown that in a vacuum environment (such as in technologies for use in space), graphite degrades as a lubricant, due to the hypoxic conditions This observation led to the hypothesis that the lubrication is due to the presence of fluids between the layers, such as air and water, which are naturally adsorbed from the environment. This hypothesis has been refuted by studies showing that air and water are not absorbed. Recent studies suggest that an effect called superlubricity can also account for graphite's lubricating properties. The use of graphite is limited by its tendency to facilitate pitting corrosion in some stainless steel, and to promote galvanic corrosion between dissimilar metals (due to its electrical conductivity). It is also corrosive to aluminium in the presence of moisture. For this reason, the US Air Force banned its use as a lubricant in aluminium aircraft, and discouraged its use in aluminium-containing automatic weapons. Even graphite pencil marks on aluminium parts may facilitate corrosion. Another high-temperature lubricant, hexagonal boron nitride, has the same molecular structure as graphite. It is sometimes called *white graphite*, due to its similar properties.

When a large number of crystallographic defects bind these planes together, graphite loses its lubrication properties and becomes what is known as pyrolytic graphite. It is also highly anisotropic, and diamagnetic, thus it will float in mid-air above a strong magnet. If it is made in a fluidized bed at 1000–1300 °C then it is isotropic turbostratic, and is used in blood contacting devices like mechanical heart valves and is called pyrolytic carbon, and is not diamagnetic. Pyrolytic graphite, and pyrolytic carbon are often confused but are very different materials.

Natural and crystalline graphites are not often used in pure form as structural materials, due to their shear-planes, brittleness and inconsistent mechanical properties.

History of Natural Graphite use

In the 4th millennium B.C., during the Neolithic Age in southeastern Europe, the Marița culture used graphite in a ceramic paint for decorating pottery.

Some time before 1565 (some sources say as early as 1500), an enormous deposit of graphite was discovered on the approach to Grey Knotts from the hamlet of Seathwaite in Borrowdale parish, Cumbria, England, which the locals found very useful for marking sheep. During the reign of Elizabeth I (1533–1603), Borrowdale graphite was used as a refractory material to line moulds for cannonballs, resulting in rounder, smoother balls that could be fired farther, contributing to the strength of the English navy. This particular deposit of graphite was extremely pure and soft, and could easily be cut into sticks. Because of its military importance, this unique mine and its production were strictly controlled by the Crown.

Other Names

Historically, graphite was called black lead or plumbago. Plumbago was commonly used in its massive mineral form. Both of these names arise from confusion with the similar-appearing lead ores, particularly galena. The Latin word for lead, *plumbum*, gave its name to the English term for this grey metallic-sheened mineral and even to the leadworts or plumbagos, plants with flowers that resemble this colour.

The term *black lead* usually refers to a powdered or processed graphite, matte black in color.

Abraham Gottlob Werner coined the name *graphite* ("writing stone") in 1789. He attempted to clear up the confusion between molybdena, plumbago and blacklead after Carl Wilhelm Scheele in 1778 proved that there are at least three different minerals. Scheele's analysis showed that the chemical compounds molybdenum sulfide (molybdenite), lead(II) sulfide (galena) and graphite were three different soft black minerals.

Uses of Natural Graphite

Natural graphite is mostly consumed for refractories, batteries, steelmaking, expanded graphite, brake linings, foundry facings and lubricants. Graphene, which occurs naturally in graphite, has unique physical properties and is among the strongest substances known. However, the process of separating it from graphite will require more technological development.

Refractories

The use of graphite as a refractory material began before 1900 with the graphite crucible used to hold molten metal; this is now a minor part of refractories. In the mid-1980s, the carbon-magnesite brick became important, and a bit later the alumina-graphite shape. As of 2017 the order of importance is: alumina-graphite shapes, carbon-magnesite brick, monolithics (gunning and ramming mixes), and then crucibles.

Crucibles began using very large flake graphite, and carbon-magnesite brick requiring not quite so large flake graphite; for these and others there is now much more flexibility in size of flake required, and amorphous graphite is no longer restricted to low-end refractories. Alumina-graphite shapes are used as continuous casting ware, such as nozzles and troughs, to convey the molten steel from ladle to mold, and carbon magnesite bricks line steel converters and electric-arc furnaces to withstand extreme temperatures. Graphite blocks are also used in parts of blast furnace linings where the high thermal conductivity of the graphite is critical. High-purity monolithics are often used as a continuous furnace lining instead of carbon-magnesite bricks.

The US and European refractories industry had a crisis in 2000–2003, with an indifferent market for steel and a declining refractory consumption per tonne of steel underlying firm buyouts and many plant closures. Many of the plant closures resulted from the acquisition of Harbison-Walker Refractories by RHI AG and some plants had their equipment auctioned off. Since much of the lost capacity was for carbon-magnesite brick, graphite consumption within the refractories area moved towards alumina-graphite shapes and monolithics, and away from brick. The major source of carbon-magnesite brick is now imports from China. Almost all of the above refractories are used to make steel and account for 75% of refractory consumption; the rest is used by a variety of industries, such as cement.

According to the USGS, US natural graphite consumption in refractories comprised 12,500 tonnes in 2010.

Batteries

The use of graphite in batteries has been increasing in the last 30 years. Natural and synthetic graphite are used to construct the anode of all major battery technologies. The lithium-ion battery utilizes roughly twice the amount of graphite than lithium carbonate.

The demand for batteries, primarily nickel-metal-hydride and lithium-ion batteries, has caused a growth in graphite demand in the late 1980s and early 1990s. This growth was driven by portable electronics, such as portable CD players and power tools. Laptops, mobile phones, tablet, and smartphone products have increased the demand for batteries. Electric vehicle batteries are anticipated to increase graphite demand. As an example, a lithium-ion battery in a fully electric Nissan Leaf contains nearly 40 kg of graphite.

Steelmaking

Natural graphite in this end use mostly goes into carbon raising in molten steel, although it can be used to lubricate the dies used to extrude hot steel. Supplying carbon raisers is very competitive, therefore subject to cut-throat pricing from alternatives such as synthetic graphite powder, petroleum coke, and other forms of carbon. A carbon raiser is added to increase the carbon content of the steel to the specified level. An estimate based on USGS US graphite consumption statistics indicates that 10,500 tonnes were used in this fashion in 2005.

Brake Linings

Natural amorphous and fine flake graphite are used in brake linings or brake shoes for heavier (nonautomotive) vehicles, and became important with the need to substitute for asbestos. This use has been important for quite some time, but nonasbestos organic (NAO) compositions are beginning to reduce graphite's market share. A brake-lining industry shake-out with some plant closures has not been beneficial, nor has an indifferent automotive market. According to the USGS, US natural graphite consumption in brake linings was 6,510 tonnes in 2005.

Foundry Facings and Lubricants

A foundry facing mold wash is a water-based paint of amorphous or fine flake graphite. Painting the inside of a mold with it and letting it dry leaves a fine graphite coat that will ease separation of the object cast after the hot metal has cooled. Graphite lubricants are specialty items for use at very high or very low temperatures, as forging die lubricant, an antiseize agent, a gear lubricant for mining machinery, and to lubricate locks. Having low-grit graphite, or even better no-grit graphite (ultra high purity), is highly desirable. It can be used as a dry powder, in water or oil, or as colloidal graphite (a permanent suspension in a liquid). An estimate based on USGS graphite consumption statistics indicates that 2,200 tonnes was used in this fashion in 2005.

Pencils

The ability to leave marks on paper and other objects gave graphite its name, given in 1789 by German mineralogist Abraham Gottlob Werner. It stems from *graphein*, meaning *to write/draw* in Ancient Greek.

Graphite pencils

From the 16th Century, pencils were made with leads of English natural graphite, but modern pencil lead is most commonly a mix of powdered graphite and clay; it was invented by Nicolas-Jacques Conté in 1795. It is chemically unrelated to the metal lead, whose ores had a similar appearance, hence the continuation of the name. Plumbago is another older term for natural graphite used for drawing, typically as a lump of the mineral without a wood casing. The term plumbago drawing is normally restricted to 17th and 18th century works, mostly portraits.

Today, pencils are still a small but significant market for natural graphite. Around 7% of the 1.1 million tonnes produced in 2011 was used to make pencils. Low-quality amorphous graphite is used and sourced mainly from China.

Other uses

Natural graphite has found uses in zinc-carbon batteries, in electric motor brushes, and various specialized applications. Graphite of various hardness or softness results in different qualities and tones when used as an artistic medium. Railroads would often mix powdered graphite with waste oil or linseed oil to create a heat-resistant protective coating for the exposed portions of a steam locomotive's boiler, such as the smokebox or lower part of the firebox.

Expanded Graphite

Expanded graphite is made by immersing natural flake graphite in a bath of chromic acid, then concentrated sulfuric acid, which forces the crystal lattice planes apart, thus expanding the graphite. The expanded graphite can be used to make graphite foil or used directly as "hot top" compound to insulate molten metal in a ladle or red-hot steel ingots and decrease heat loss, or as firestops fitted around a fire door or in sheet metal collars surrounding plastic pipe (during a fire, the graphite expands and chars to resist fire penetration and spread), or to make high-performance gasket material for high-temperature use. After being made into graphite foil, the foil is machined and assembled into the bipolar plates in fuel cells. The foil is made into heat sinks for laptop computers which keeps them cool while saving weight, and is made into a foil laminate that can be used in valve packings or made into gaskets. Old-style packings are now a minor member of this grouping: fine flake graphite in oils or greases for uses requiring heat resistance. A GAN estimate of current US natural graphite consumption in this end use is 7,500 tonnes.

Intercalated Graphite

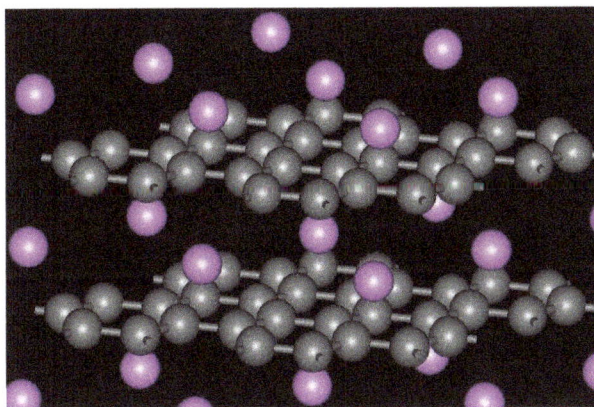

Structure of CaC_6

Graphite forms intercalation compounds with some metals and small molecules. In these compounds, the host molecule or atom gets "sandwiched" between the graphite layers, resulting in a type of compounds with variable stoichiometry. A prominent example of an intercalation compound is potassium graphite, denoted by the formula KC_8. Graphite intercalation compounds are superconductors. The highest transition temperature (by June 2009) $T_c = 11.5$ K is achieved in CaC_6, and it further increases under applied pressure (15.1 K at 8 GPa).

Uses of Synthetic Graphite

Invention of a Process to Produce Synthetic Graphite

In 1893 Charles Street of Le Carbone discovered a process for making artificial graphite. Another process to make synthetic graphite was invented accidentally by Edward Goodrich Acheson (1856–1931). In the mid-1890s, Acheson discovered that overheating carborundum produced almost pure graphite. While studying the effects of high temperature on carborundum, he had found that silicon vaporizes at about 4,150 °C (7,500 °F), leaving the carbon behind in graphitic carbon. This graphite was another major discovery for him, and it became extremely valuable and helpful as a lubricant.

In 1896 Acheson received a patent for his method of synthesizing graphite, and in 1897 started commercial production. The Acheson Graphite Co. was formed in 1899.

Scientific Research

Highly oriented pyrolytic graphite (HOPG) is the highest-quality synthetic form of graphite. It is used in scientific research, in particular, as a length standard for scanner calibration of scanning probe microscope.

Electrodes

Graphite electrodes carry the electricity that melts scrap iron and steel, and sometimes direct-reduced iron (DRI), in electric arc furnaces, which are the vast majority of steel furnaces. They are made from petroleum coke after it is mixed with coal tar pitch. They are then extruded and shaped, baked to carbonize the binder (pitch), and finally graphitized by heating it to temperatures approaching 3000 °C, at which the carbon atoms arrange into graphite. They can vary in size up to 11 feet long and 30 inches in diameter. An increasing proportion of global steel is made using electric arc furnaces, and the electric arc furnace itself is getting more efficient, making more steel per tonne of electrode. An estimate based on USGS data indicates that graphite electrode consumption was 197,000 tonnes in 2005.

Electrolytic aluminium smelting also uses graphitic carbon electrodes. On a much smaller scale, synthetic graphite electrodes are used in electrical discharge machining (EDM), commonly to make injection molds for plastics.

Powder and Scrap

The powder is made by heating powdered petroleum coke above the temperature of graphitization, sometimes with minor modifications. The graphite scrap comes from pieces of unusable electrode material (in the manufacturing stage or after use) and lathe turnings, usually after crushing and sizing. Most synthetic graphite powder goes to carbon raising in steel (competing with natural graphite), with some used in batteries and brake linings. According to the USGS, US synthetic graphite powder and scrap production was 95,000 tonnes in 2001 (latest data).

Neutron Moderator

Special grades of synthetic graphite, such as Gilsocarbon, also find use as a matrix and neutron moderator within nuclear reactors. Its low neutron cross-section also recommends it for use in proposed fusion reactors. Care must be taken that reactor-grade graphite is free of neutron absorbing materials such as boron, widely used as the seed electrode in commercial graphite deposition systems—this caused the failure of the Germans' World War II graphite-based nuclear reactors. Since they could not isolate the difficulty they were forced to use far more expensive heavy water moderators. Graphite used for nuclear reactors is often referred to as nuclear graphite.

Other uses

Graphite (carbon) fiber and carbon nanotubes are also used in carbon fiber reinforced plastics, and in heat-resistant composites such as reinforced carbon-carbon (RCC). Commercial structures made from carbon fiber graphite composites include fishing rods, golf club shafts, bicycle frames, sports car body panels, the fuselage of the Boeing 787 Dreamliner and poolcue sticks and have been successfully employed in reinforced concrete, The mechanical properties of carbon fiber graphite-reinforced plastic composites and grey cast iron are strongly influenced by the role of graphite in these materials. In this context, the term "(100%) graphite" is often loosely used to refer to a pure mixture of carbon reinforcement and resin, while the term "composite" is used for composite materials with additional ingredients.

Modern smokeless powder is coated in graphite to prevent the buildup of static charge.

Graphite has been used in at least three radar absorbent materials. It was mixed with rubber in Sumpf and Schornsteinfeger, which were used on U-boatsnorkels to reduce their radar cross section. It was also used in tiles on early F-117 Nighthawk (1983)s.

Graphite Mining, Beneficiation, and Milling

Graphite is mined by both open pit and underground methods. Graphite usually needs beneficiation. This may be carried out by hand-picking the pieces of gangue (rock) and hand-screening the product or by crushing the rock and floating out the graphite. Beneficiation by flotation encounters the difficulty that graphite is very soft and "marks" (coats) the particles of gangue. This makes the "marked" gangue particles float off with the graphite, yielding impure concentrate. There are two ways of obtaining a commercial concentrate or product: repeated regrinding and floating (up to seven times) to purify the concentrate, or by acid leaching (dissolving) the gangue with hydrofluoric acid (for a silicate gangue) or hydrochloric acid (for a carbonate gangue).

In milling, the incoming graphite products and concentrates can be ground before being classified (sized or screened), with the coarser flake size fractions (below 8 mesh, 8–20 mesh, 20–50 mesh) carefully preserved, and then the carbon contents are determined. Some standard blends can be prepared from the different fractions, each with a certain flake size distribution and carbon content. Custom blends can also be made for individual customers who want a certain flake size distribution and carbon content. If flake size is unimportant, the concentrate can be ground more freely. Typical end products include a fine powder for use as a slurry in oil drilling and coatings for foundry molds, carbon raiser in the steel industry (Synthetic graphite powder and powdered petroleum coke can also be used as carbon raiser). Environmental impacts from graphite mills consist of air pollution including fine particulate exposure of workers and also soil contamination from powder spillages leading to heavy metal contamination of soil.

Occupational Safety

People can be exposed to graphite in the workplace by breathing it in, skin contact, and eye contact.

United States

The Occupational Safety and Health Administration (OSHA) has set the legal limit (permissible exposure limit) for graphite exposure in the workplace as a time weighted average (TWA) of 15 million particles per cubic foot (1.5 mg/m^3) over an 8-hour workday. The National Institute for Occupational Safety and Health (NIOSH) has set a recommended exposure limit (REL) of TWA 2.5 mg/m^3 respirable dust over an 8-hour workday. At levels of 1250 mg/m^3, graphite is immediately dangerous to life and health.

Graphite Recycling

The most common way of recycling graphite occurs when synthetic graphite electrodes are either manufactured and pieces are cut off or lathe turnings are discarded, or the electrode (or other) are used all the way down to the electrode holder. A new electrode replaces the old one, but a sizeable piece of the old electrode remains. This is crushed and sized, and the resulting graphite powder is mostly used to raise the carbon content of molten steel. Graphite-containing refractories are sometimes also recycled, but often not because of their graphite: the largest-volume items, such as carbon-magnesite bricks that contain only 15–25% graphite, usually contain too little graphite. However, some recycled carbon-magnesite brick is used as the basis for furnace-repair materials, and also crushed carbon-magnesite brick is used in slag conditioners. While crucibles have a high graphite content, the volume of crucibles used and then recycled is very small.

A high-quality flake graphite product that closely resembles natural flake graphite can be made from steelmaking kish. Kish is a large-volume near-molten waste skimmed from the molten iron feed to a basic oxygen furnace, and consists of a mix of graphite (precipitated out of the supersaturated iron), lime-rich slag, and some iron. The iron is recycled on site, leaving a mixture of graphite and slag. The best recovery process uses hydraulic classification (which utilizes a flow of water to separate minerals by specific gravity: graphite is light and settles nearly last) to get a 70% graphite rough concentrate. Leaching this concentrate with hydrochloric acid gives a 95% graphite product with a flake size ranging from 10 mesh down.

Structure of Graphite

- Other forms of Carbon such as graphite and fullerene are also covalent bonded but the structures are entirely different.

- Graphite has a layered structure where in each layer, carbon atoms are sp² hybridized and they make a hexagonal pattern. However, the bonding between individual layers is Van der Walls type of bonding. That is why Graphite is a soft material and is used as a lubricant.

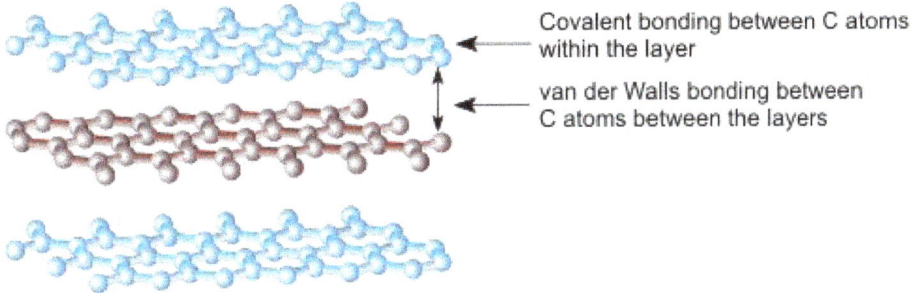

Covalent bonding between C atoms within the layer

van der Walls bonding between C atoms between the layers

Structure of Graphite

Ionically Bonded Ceramic Structures

- Most of the ceramic materials are compounds with anions and cations with different electronegativities. Hence, when these ions are brought together, they form a very strong ionic bond.

- Typically, since anions are bigger in size than cations, anions tend to form the base lattice and cations fill in the interstices.

- However, it is not so simple. As there is an involvement of two different types of ions to form a crystal structure, there are certain rules or say guidelines which need to be followed to give rise to a stable crystal structure. These rules are called Pauling's rules.

- Based on these rules, typically ceramic structures are based on anions forming the base lattice and cations occupying the interstices in them. Fortunately, most ceramic compounds are completely or partially ionically bonded and happen to be based on either of FCC or HCP packing of anions. As a result, we can categorize the structures of most ceramic materials into following categories

  o   Compounds based on cubic closed packing (CCP or FCC) of ions

  o   Compounds based on hexagonal closed packing (HCP) of ions

  o   Other structures with some deviations from above two.

Pauling's Rules

Pauling's rules are five rules published by Linus Pauling in 1929 for predicting and rationalizing the crystal structures of ionic compounds.

First Rule: The Radius Ratio Rule

For typical ionic solids, the cations are smaller than the anions, and each cation is surrounded by coordinated anions which form a polyhedron. The sum of the ionic radii determines the cation-anion distance, while the cation-anion radius ratio r_+/r_- (or r_c/r_a) determines the coordination number (C.N.) of the cation, as well as the shape of the coordinated polyhedron of anions.

For the coordination numbers and corresponding polyhedra in the table below, Pauling mathematically derived the *minimum* radius ratio for which the cation is in contact with the given number of anions (considering the ions as rigid spheres). If the cation is smaller, it will not be in contact with the anions which results in instability leading to a lower coordination number.

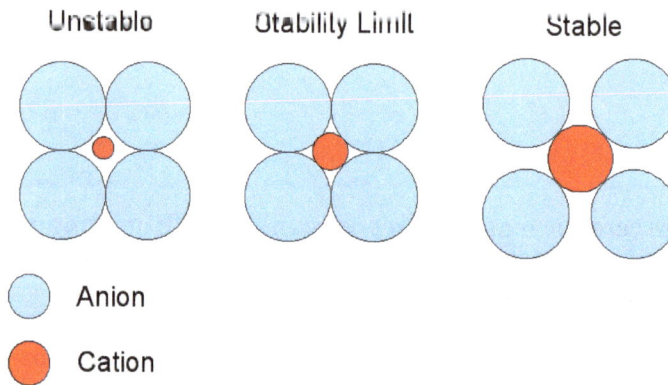

Unstable   Stability Limit   Stable

○ Anion

● Cation

Critical Radius Ratio. This diagram is for coordination number six: 4 anions in the plane shown, 1 above the plane and 1 below. The stability limit is at $r_C/r_A = 0.414$

C.N.	Polyhedron and minimum radius ratio for each coordination number	
C.N.	**Polyhedron**	**Radius ratio**
3	triangular	0.155
4	tetrahedron	0.225
6	octahedron	0.414
7	capped octahedron	0.592
8	square antiprism (anticube)	0.645
8	cube	0.732
9	triaugmented triangular prism	0.732
12	cuboctahedron	1.00

The three diagrams at right correspond to octahedral coordination with a coordination number of six: four anions in the plane of the diagrams, and two (not shown) above and below this plane. The central diagram shows the minimal radius ratio. The cation and any two anions form a right triangle, with $2r_- = \sqrt{2}(r_- + r_+)$, or $\sqrt{2}r_- = r_- + r_+$. Then

$r_+ = (\sqrt{2} - 1)r_- = 0.414 r_-$. Similar geometrical proofs yield the minimum radius ratios for the highly symmetrical cases C.N. = 3, 4 and 8.

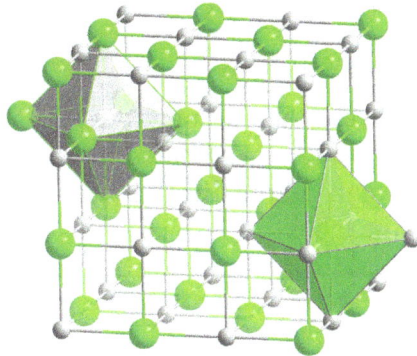

The NaCl crystal structure. Each Na atom has six nearest neighbors, with octahedral geometry.

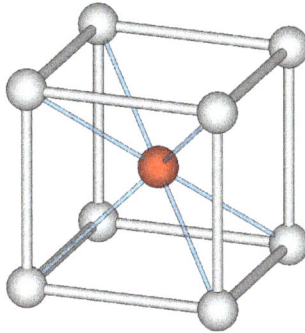

The CsCl unit cell. Each Cs atom has eight nearest neighbors, with cubic geometry.

For C.N. = 6 and a radius ratio greater than the minimum, the crystal is more stable since the cation is still in contact with six anions, but the anions are further from each other so that their mutual repulsion is reduced. An octahedron may then form with a radius ratio greater than or equal to .414, but as the ratio rises above .732, a cubic geometry becomes more stable. This explains why Na^+ in NaCl with a radius ratio of 0.55 has octahedral co-ordination, whereas Cs^+ in CsCl with a radius ratio of 0.93 has cubic coordination.

If the radius ratio is less than the minimum, two anions will tend to depart and the remaining four will rearrange into a tetrahedral geometry where they are all in contact with the cation.

The radius ratio rules are a first approximation which have some success in predicting coordination numbers, but many exceptions do exist.

Second Rule: The Electrostatic Valence Rule

For a given cation, Pauling defined the *electrostatic bond strength* to each coordinated anion as $s = \dfrac{z}{v}$, where z is the cation charge and v is the cation coordination number.

A stable ionic structure is arranged to preserve *local electroneutrality,* so that the sum of the strengths of the electrostatic bonds to an anion equals the charge on that anion.

$$\xi = \sum_i s_i$$

where ξ is the anion charge and the summation is over the adjacent cations. For simple solids, the s_i are equal for all cations coordinated to a given anion, so that the anion coordination number is the anion charge divided by each electrostatic bond strength. Some examples are given in the table.

Cations with oxide O^{2-} ion				
Cation	Radius ratio	Cation C.N.	Electrostatic bond strength	Anion C.N.
Li^+	0.34	4	0.25	8
Mg^{2+}	0.47	6	0.33	6
Sc^{3+}	0.60	6	0.5	4

Pauling showed that this rule is useful in limiting the possible structures to consider for more complex crystals such as the aluminosilicate mineral orthoclase, $KAlSi_3O_8$, with three different cations.

Third Rule: Sharing of Polyhedron Corners, Edges and Faces

The sharing of edges and particularly faces by two anion polyhedra decreases the stability of an ionic structure. Sharing of corners does not decrease stability as much, so (for example) octahedra may share corners with one another.

The decrease in stability is due to the fact that sharing edges and faces places cations in closer proximity to each other, so that cation-cation electrostatic repulsion is increased. The effect is largest for cations with high charge and low C.N. (especially when r+/r- approaches the lower limit of the polyhedral stability).

As one example, Pauling considered the three mineral forms of titanium dioxide, each with a coordination number of 6 for the Ti^{4+} cations. The most stable (and most abundant) form is rutile, in which the coordination octahedra are arranged so that each one shares only two edges (and no faces) with adjoining octahedra. The other two, less stable, forms are brookite and anatase, in which each octahedron shares three and four edges respectively with adjoining octahedra.

Fourth Rule: Crystals Containing Different Cations

In a crystal containing different cations, those of high valency and small coordination number tend not to share polyhedron elements with one another. This rule tends to increase the distance between highly charged cations, so as to reduce the electrostatic repulsion between them.

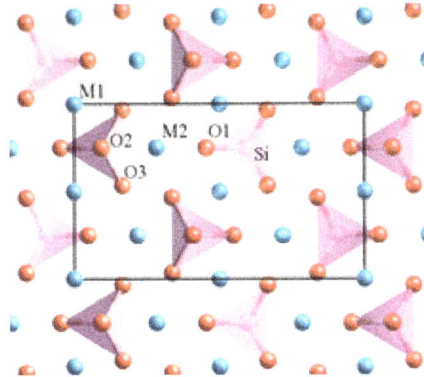

Structure of olivine. M (Mg or Fe) = blue spheres, Si = pink tetrahedra, O = red spheres.

One of Pauling's examples is olivine, M_2SiO_4, where M is a mixture of Mg^{2+} at some sites and Fe^{2+} at others. The structure contains distinct SiO_4 tetrahedra which do not share any oxygens (at corners, edges or faces) with each other. The lower-valence Mg^{2+} and Fe^{2+} cations are surrounded by polyhedra which do share oxygens.

Fifth Rule: the Rule of Parsimony

The number of essentially different kinds of constituents in a crystal tends to be small. The repeating units will tend to be identical because each atom in the structure is most stable in a specific environment. There may be two or three types of polyhedra, such as tetrahedra or octahedra, but there will not be many different types.

Bond Strength

In chemistry, bond strength is the degree to which each atom joined to another in a chemical bond contributes to the valency of this other atom. Bond strength is intimately linked to bond order and can be quantified by:

- bond energy: requires lengthy calculations, even for the simplest bonds.

- bond-dissociation energy

- (relaxed) force constant

Another criterion of bond strength is the qualitative relation between bond energies and the overlap of atomic orbitals of the bonds (Pauling and Mulliken). The more these overlap, the more the bonding electrons are to be found between the nuclei and hence stronger will be the bond. Overlap is necessary for the formation of molecular orbitals. This overlap can be calculated and is called the overlap integral.

- Bond strength is a useful parameter to determine whether the derived structure is correct or not, at least whether the charge is neutral and stoichiomteric or not. Bond strength of an ion is defined as the ratio of the valence of an ion to its co-ordination, *i.e.*

$$Bond\ strength = \frac{Valenece\ of\ the\ ion}{Coordination\ nunber\ of\ ion}$$

- In a stoichiometric and charge neutral solid, the bond strengths of cations must be equal to those of anions. Alternatively, you can work out bond strength of one ion and from this, you can work out the valence of other ion which should match what is needed to maintain the stoichiometry and most cases, the common valence state.

Other Cubic Structures

There are a few structures, which appear as if they are based on cubic closed packing of anions. However the actual structure is rather different and many of these structures are merely based on the cubic packing of anions.

Perovskite (ABO_3) Structure

- ABO_3 type compounds

- Examples are many titanates like $BaTiO_3$, $SrTiO_3$, $PbTiO_3$ etc. which happen to be technologically very useful compounds.

- In ABO_3 structured compounds, A ion is twelve fold coordinated by oxygen (like a dodecahedra) and B ion is octahedrally coordinated by oxygen ions.

- Oxygen atoms form an FCC-like (not FCC) cell with atoms missing from the corners which are occupied by A atoms.

- Bond strength check:

Cation: Ba: $2/12 = 1/6$ and Ti: $4/6 = 2/3$

Oxygen valence = $\frac{1}{6}$ x Coordination number by Ba + $\frac{2}{3}$ x coordination number by Ti .

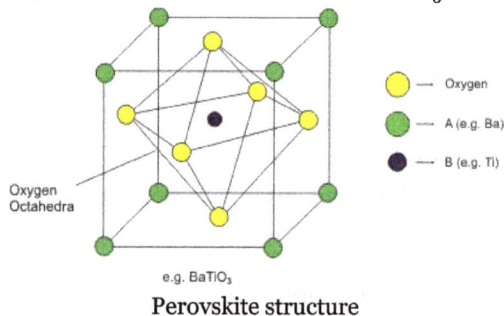

Perovskite structure

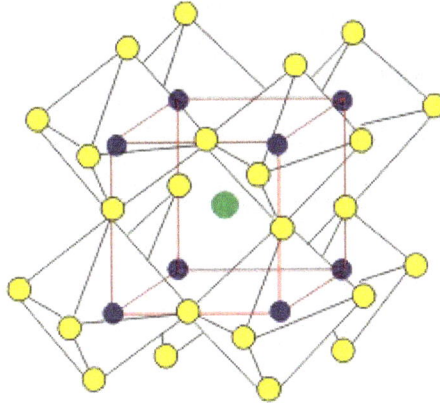

Polyhedra model of perovskite structure

- Lattice type: Primitive Cubic (NOT FCC!)

- Motif: A ion - 0 0 0, B ion – ½ ½ ½, O ion - ½ ½ 0, 0 ½ ½, ½ 0 ½

- One Formula unit per unit cell

- Coordination

   o   B cation is surrounded by oxygen octahedra which share corners.

   o   A cation is surrounded by oxygen dodecahedra which touch faces of oc-
       tahedra.

- An important parameters about perovskites is the their "Tolerance Factor (t)"
  which is defined as

$$t = \frac{r_A + r_0}{\sqrt{2}(r_B + r_0)}$$

- This is derived from the geometry of a cube in which the atoms are of such sizes
  that they touch each other and hence, the face diagonal of the unit cell would be
  $\sqrt{2}$ times the unit-cell length, as result $t = 1$ for a perfect cubic perovskite.

- However, due to variations in ionic radii of various ions, many perovskites
  show deviations from $t = 1$ and may not even have a cubic structure. Deviations
  from $t = 1$ signify the level of lattice distortion.

- For example, $BaTiO_3$ has cubic structure only above ~120°C while it is tetragonal
  at room temperature and further adopts orthorhombic and rhombohedral
  structure if cooled below RT.

- Perovskites can also have various combinations of ionic valence such as

   o   e.g. $A^{2+}B^{4+}O_4$, $BaTiO_3$, $PbTiO_3$, $CaTiO_3$, $SrTiO_3$ etc.

o   e.g. $A^{3+}B^{3+}O_4$, $LaAlO_3$, $LaGaO_3$, $BiFeO_3$ etc.

o   Mixed Perovskites:

□ $A^{2+}(B^{2+}_{1/3}B^{5+}_{2/3})O_3$ eg. $Pb(Mg_{1/3}Nb_{2/3})O_3$

□ $A^{2+}(B^{3+}_{1/2}B^{5+}_{1/2})O_3$ eg. $Pb(Sc_{1/2}Ta_{1/2})O_3$

ReO_3 Structure

- Stoichiometry : MX_3

- Lattice type: Primitive cubic

- Atomic Positions: M- 0 0 0; X - ½ 0 0, 0 ½ 0, 0 0 ½

- Coordination Numbers

M	CN= 6	Octahedral coordination
X	CN = 2	Linear coordination

- Can be visualized as perovskite ABO_3 structure with empty B-sites

- Representative Oxides

- ReO_3, UO_3, WO_3

- Used for gas sensing and electrochromic applications

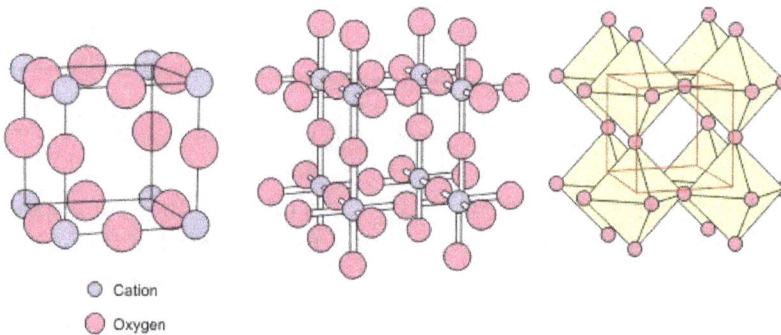

○ Cation
● Oxygen

ReO_3 structure and polyhedra model

CsCl Structure

- MX type compounds, parent compound being CsCl.

- Examples: Halides such as CSCl, AgI, AgBr etc.

- Radius ratio governs cubic co-ordination of both cations and anions.

- Lattice type: Primitive cubic lattice.

- Motif: Anions (X): o o o, Cations (M): ½ ½ ½

- One formula unit per unit cell.

- Primitive cubic lattice

- Motif C_s $\frac{1}{2}$ $\frac{1}{2}$ $\frac{1}{2}$
  C_l 0 0 0

- One formula unit per unit cell

Anion
Cation

(a) CsCl structure (b) Ball-stick model

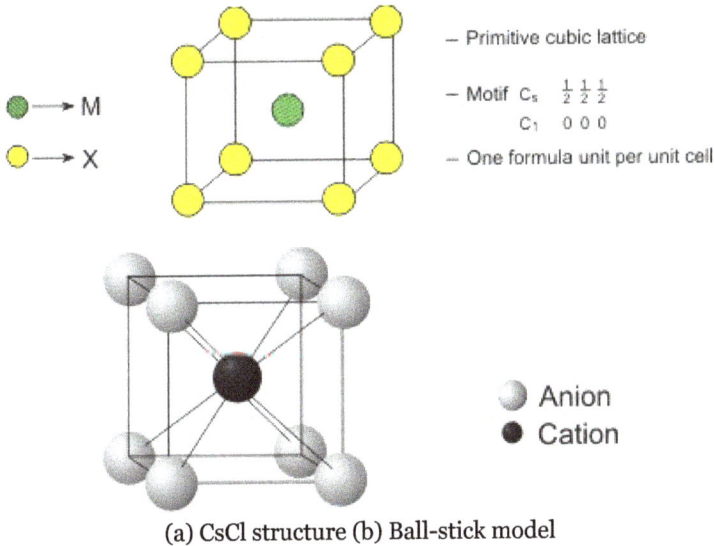

Orthogonal Structures

- Many superconductors follow the structures which are perovskite based i.e., the structure contains the perovskite structured units stacked along c-axis or -direction in most cases. The examples are superconductors such as $YBa_2Cu_3O_7$, ferroelectrics such as $Bi_4Ti_3O_{12}$ etc. In some other compounds such as La-Sr-Cu-O, the structure is composed of alternating perovskite and rocksalt structure units. Such a representation makes it easy to understand them.

- Here we will take examples of Y-Ba-Cu-O and La-Sr-Cu-O and discuss them very briefly.

Yttrium Barium Copper Oxide or YBCO ($YBa_2Cu_3O_7$)

- Parent compound is $Y_3Cu^{3+}_3O_9$ which also contains perovskite units.

- Doping of Y by Ba leads to structure modification as well as reduction of Cu^{3+} to Cu^{2+} state and thus resulting in the reduction in the number of required oxygen ions and hence creates oxygen vacancies in the structure. This gives a transition temperature of ~92 K below which the compound has zero electrical resistance i.e. is a superconductor.

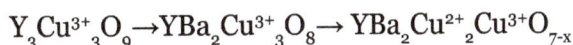

$$Y_3Cu^{3+}_3O_9 \rightarrow YBa_2Cu^{3+}_3O_8 \rightarrow YBa_2Cu^{2+}_2Cu^{3+}O_{7-x}$$

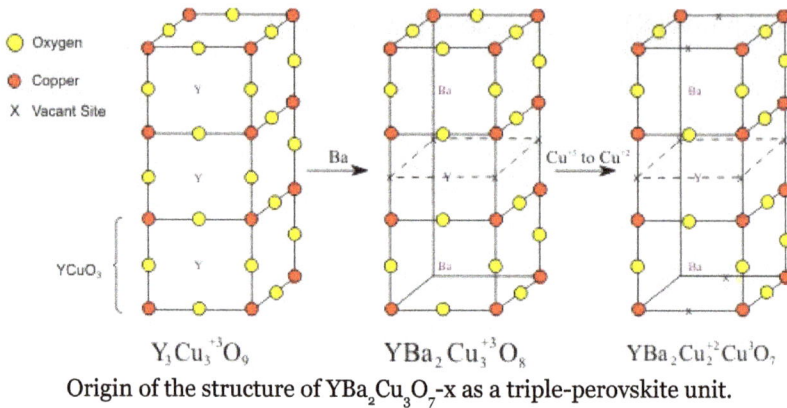

$$Y_3Cu_3^{+3}O_9 \qquad YBa_2Cu_3^{+3}O_8 \qquad YBa_2Cu_2^{+2}Cu^1O_7$$

Origin of the structure of $YBa_2Cu_3O_7$-x as a triple-perovskite unit.

- Here Cu coordination is of interest:

    o Cu^{2+} atoms have four-fold coordination along Cu-O chains.

    o Cu^{3+} atoms have five-fold coordination in the Cu-O planes.

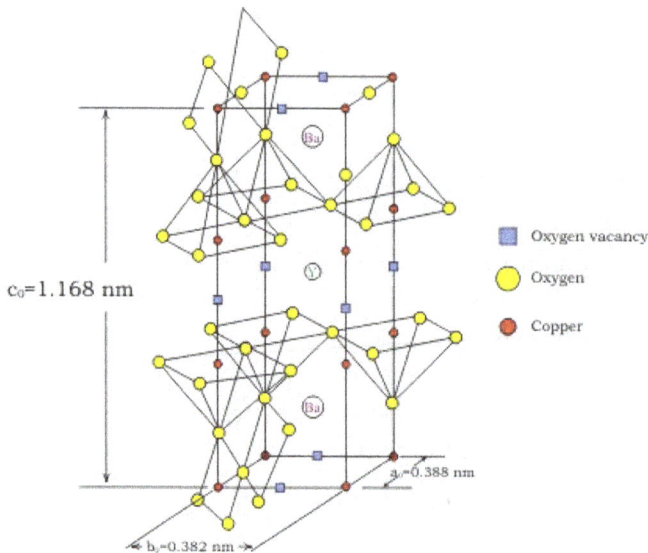

Atomic coordination in YBCO

Lanthanum Strontium Copper Oxide $La_{2-x}Sr_xCuO_4$

- Parent compound La_2CuO_4 is actually a mixture of one Rocksalt structured compound, LaO and one perovskite structured compound, $LaCuO_3$ and can also be written as $LaO.LaCuO_3$.

- The structure shows a layered structure with layers stacked as A_4O-AO_4-A_4O as shown below where A is La.

- Substitution of La by Sr results in the compound $La_{2-x}Sr_xCuO_4$ turning into a superconductor with a Tc ~ 35K.

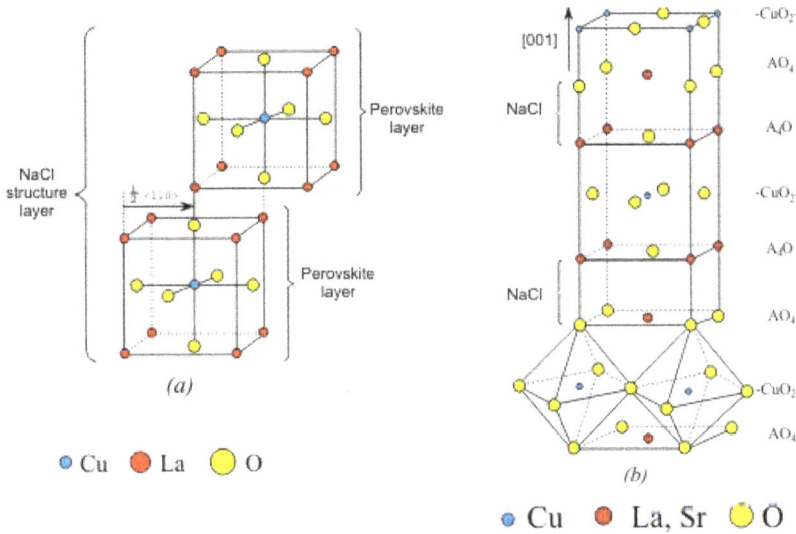

a) Origin of $La_{2-x}Sr_xCuO_4$ structure, shown in (b) as two perovskite and cells

Structures Based on HCP Packing of Ions

Similar to FCC packing of anions, many ceramic structures are also based on another type of closed packing of anions i.e. hexagonal closed packed (HCP). In this category we will look at the following structures:

- Wurzite structured compounds

- Corundum structured compounds

- Ilmenite structure compounds

- Lithium niobate structured compounds and

- Rutile structure

Wurtzite (MX) Structured Compounds

- Compounds with $M^{2+}X^{2-}$ stoichiometry

- Examples are the polymorphs of Sphalerite structured compounds such as ZnS ZnO, SiC.

- Co-ordination of both anions and cations is 4, as in Sphalerite structured compounds.

- Anions form an HCP lattice with ½ of the tetrahedral sites occupied by cations.

- The only difference to Sphalerite structure is that here anions pack in the form of ABCABC... stacking.

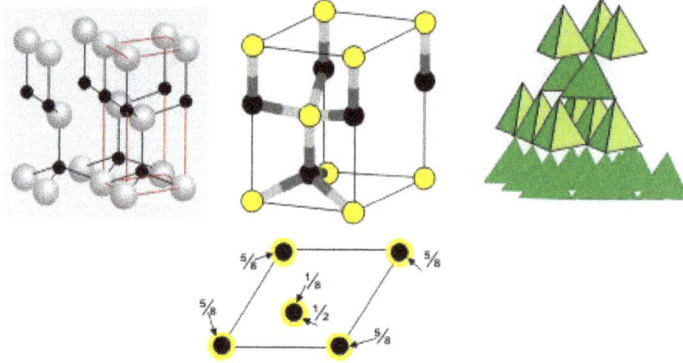

Wurtzite structure and polyhedral model

- As you can notice, all the tetrahedrons point in one direction i.e. along the c-axis of the unit-cell and they share the corners.

- Lattice type: Primitive, HCP

- Motif: M: o o o and ; X: and

- The filling of structure can be seen below.

Wurtzite (e.g. ZnO)

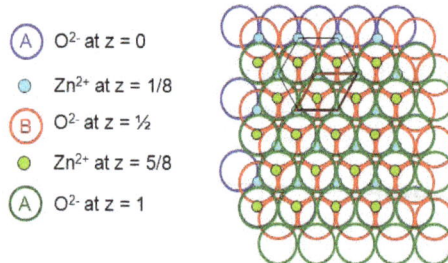

- (A) O^{2-} at z = 0
- ○ Zn^{2+} at z = 1/8
- (B) O^{2-} at z = ½
- ○ Zn^{2+} at z = 5/8
- (A) O^{2-} at z = 1

Layer by layer filling in Wurtzite

Corundum (Al_2O_3) Structured Compounds

- M_2X_3 type of compounds

- α - Alumina or Sapphire (Al_2O_3) is the parent compound.

  o   Other examples are compounds like Cr_2O_3, Fe_2O_3

- Anions form an HCP lattice

- Two-third of octahedral voids are occupied by the cations to maintain the stoichiometry.

- Coordination numbers: M: 6, X: 4.

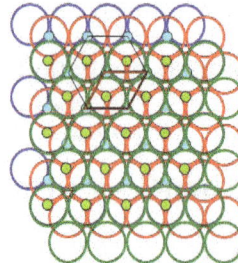

- This arrangement preserves the charge neutrality as you can also verify using bond strength formula.

- This can be best viewed when we look at the basal plane of (0001)-plane of the unit-cell and start filling the interstices.

Filling of Spheres: Corundum

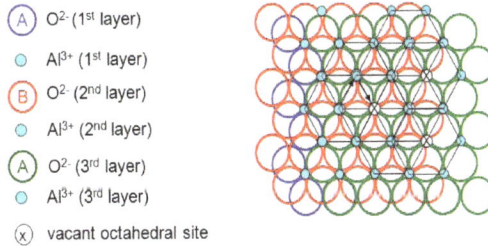

- (A) O^{2-} (1st layer)
- (○) Al^{3+} (1st layer)
- (B) O^{2-} (2nd layer)
- (○) Al^{3+} (2nd layer)
- (A) O^{2-} (3rd layer)
- (○) Al^{3+} (3rd layer)
- (x) vacant octahedral site

Whole structure consists of 6 layers of oxygen

Layered filling of Corundum

- One unit-cell consists of six layers of oxygen ions.

- A side view of the structure on plane can be seen below where you can see columns of cations along the c-axis with $\frac{2}{3}$ rd filling of octahedral sites.

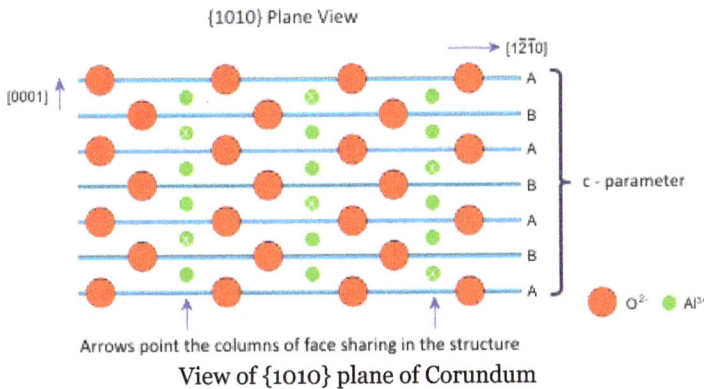

{1010} Plane View

Arrows point the columns of face sharing in the structure

View of {1010} plane of Corundum

Ilmenite Structure

The stoichiomteric formula is ABO_3 (different to perovskite ABO_3)

- The parent compound is $FeTiO_3$.

- Other compounds which follow this structure are $CdTiO_3$, $CoTiO_3$, $CrRhO_3$, $FeRhO_3$, $FeVO_3$, $LiNbO_3$, $MgGeO_3$, $MgTiO_3$.

- This structure is very similar to Corundum or $\alpha - Al_2O_3$.

- Imagine the Corundum structure and replace Al atoms in the octahedral sites in one (0001) layer *i.e.* half of the total aluminum atoms by Fe and the remaining half in the next layer by Ti atoms in the octahedral sites and continue this order of substitution along the *c*-axis of the unit-cell.

- Hence, the atomic arrangement is similar to Al_2O_3 except with alternate layers of Fe and Ti in place of Al.

- Coordination numbers: both Fe and Ti remain octahedrally coordinated while O is coordinated by 4 cations i.e. 2 Fe and 2 Ti.

- Bond strength rule gives correct oxygen valence:

$$\frac{2(valence\ of\ Fe)}{6(CN\ of\ Fe)} \times 2(CN\ of\ 0\ by\ Fe) + \frac{4(valence\ of\ Ti)}{6(CN\ of\ Ti)} \times 2(CN\ of\ 0\ by\ Ti) = 2 = Oxygen\ valence$$

Filling of Spheres: Ilmenite

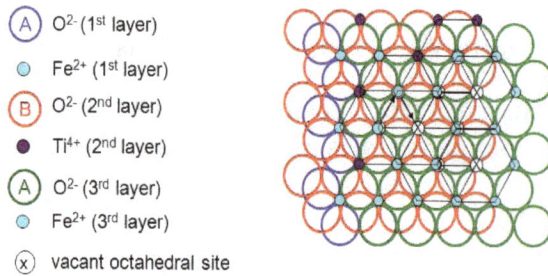

(A) O^{2-} (1st layer)
○ Fe^{2+} (1st layer)
(B) O^{2-} (2nd layer)
● Ti^{4+} (2nd layer)
(A) O^{2-} (3rd layer)
○ Fe^{2+} (3rd layer)
(x) vacant octahedral site

Whole structure consists of 6 layers of oxygen

Layered filling of Ilmenite

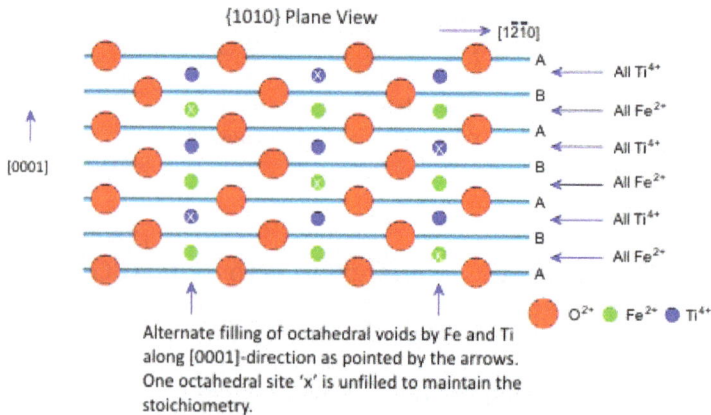

Alternate filling of octahedral voids by Fe and Ti along [0001]-direction as pointed by the arrows. One octahedral site 'x' is unfilled to maintain the stoichiometry.

Ilmenite structure on {10-10} plane

- One unit-cell consists of six layers of oxygen ions.

- A side view of the structure on {10-10} plane, as shown below, shows the columns of cations along the c-axis with $\tfrac{2}{3}^{rd}$ filling of octahedral sites which are alternately filled by Fe and Ti ions and then followed by a vacant site.

Lithium Niobate Structure

- Structure is similar to Al_2O_3 except that Al sub-lattice is substituted in an ordered manner by Li and Nb ions in the same layer unlike in alternating layer in Fe_2O_3.

- The parent compound $LiNbO_3$ is ferroelectric in nature and hence, is technologically important.

- $LiNbO_3$ also has highly anisotropic refractive index and it shows birefringence which is changeable by electric field.

- Such materials are used in electro-optic devices.

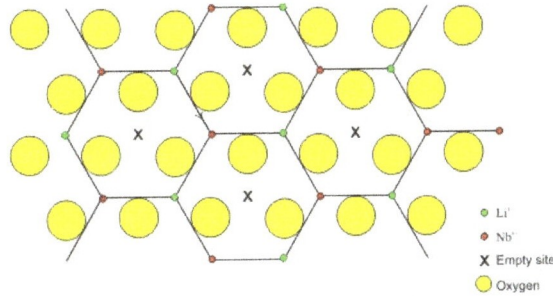

Atomic arrangement of a layer in $LiNbO_3$ structure

{1010} Plane View

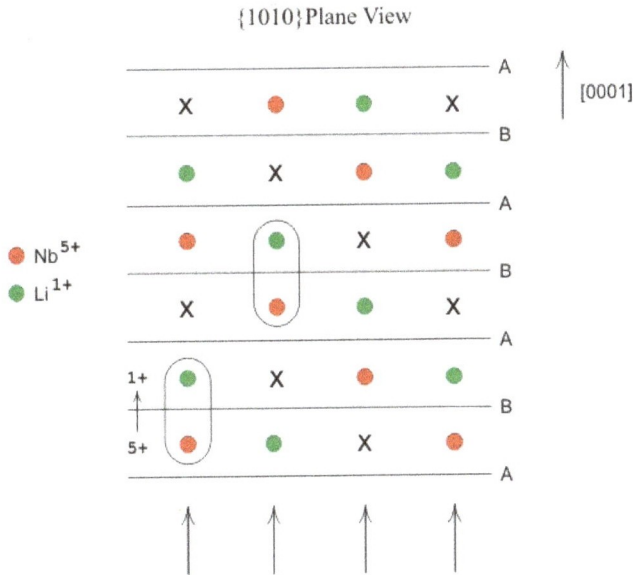

Structure on {10-10} plane in $LiNbO_3$

Rutile Structure

- Polymorph of titanium di-oxide or TiO_2

- Other forms are Anatase and Brookite.

- It is formed by quasi-HCP packing of anions.

- Half of the octahedral sites are filled by cations.

- The resulting structure has a tetragonal crystal structure due to a slight distortion in the lattice.

- Anisotropic diffusion properties of cations are found in TiO_2.

- Materials shows large and anisotropic refractive index and high birefringence.

- TiO_2 is often used as pigments and is non-toxic.

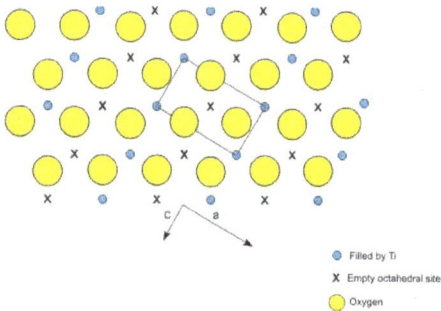

Structure of a layer of oxygen and Titanium in Rutile

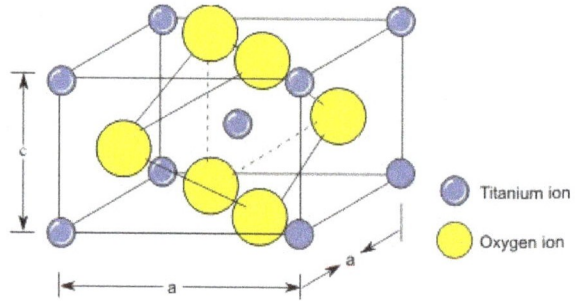

Unit-cell of Rutile

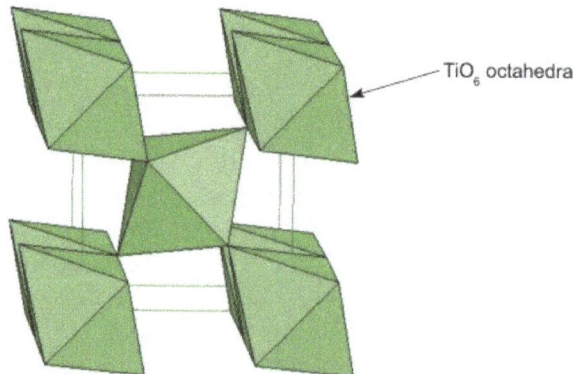

Polyhedral model of Rutile

References

- Lavrakas, Vasilis (1957). "Textbook errors: Guest column. XII: The lubricating properties of graphite". Journal of Chemical Education. 34 (5): 240. Bibcode:1957JChEd..34..240L. doi:10.1021/ed034p240

- Black, J. T.; Kohser, R. A. (2012). DeGarmo's materials and processes in manufacturing. Wiley. p. 226. ISBN 978-0-470-92467-9

- Garvie, R. C.; Hannink, R. H.; Pascoe, R. T. (1975). "Ceramic steel?". Nature. 258 (5537): 703–704. Bibcode:1975Natur.258..703G. doi:10.1038/258703a0

- Mississippi Valley Archaeological Center, Ceramic Analysis Archived June 3, 2012, at the Wayback Machine., Retrieved 04-11-12

- Carter, C. B.; Norton, M. G. (2007). Ceramic materials: Science and engineering. Springer. pp. 3 & 4. ISBN 978-0-387-46271-4

- Sundquist, J. J.; Lin, C. C. (1981). "Electronic structure of the F centre in a sodium fluoride crystal". Journal of Physics C: Solid State Physics. 14 (32): 4797–4805. Bibcode:1981JPhC...14.4797S. doi:10.1088/0022-3719/14/32/016

- Abrahams, S. C.; Bernstein, J. L. (1965). "Accuracy of an automatic diffractometer. Measurement of the sodium chloride structure factors". Acta Crystallogr. 18 (5): 926–932. doi:10.1107/S0365110X65002244

- Prince, E., ed. (2006). International Tables for Crystallography. International Union of Crystallography. ISBN 978-1-4020-4969-9. doi:10.1107/97809553602060000001

- Parhami, B.; Kwai, Ding-Ming (2001), "A unified formulation of honeycomb and diamond networks", IEEE Transactions on Parallel and Distributed Systems, 12 (1): 74–80, doi:10.1109/71.899940

- Yen, Bing; Schwickert, Birgit (2004). Origin of low-friction behavior in graphite investigated by surface x-ray diffraction, SLAC-PUB-10429 (PDF) (Report). Retrieved March 15, 2013

- Greenwood, Norman N.; Earnshaw, Alan (1997). Chemistry of the Elements (2nd ed.). Butterworth-Heinemann. ISBN 0-08-037941-9

- Kao, W.; Peretti, E. (1970). "The ternary subsystem Sn4As3-SnAs-SnTe". Journal of the Less Common Metals. 22: 39–50. doi:10.1016/0022-5088(70)90174-8

- Evans, John W. (1908). "V.— the Meanings and Synonyms of Plumbago". Transactions of the Philological Society. 26 (2): 133–179. doi:10.1111/j.1467-968X.1908.tb00513.x

- Kobashi, Koji (2005), "2.1 Structure of diamond", Diamond films: chemical vapor deposition for oriented and heteroepitaxial growth, Elsevier, p. 9, ISBN 978-0-08-044723-0

- Hanaor, D.; Michelazzi, M.; Leonelli, C.; Sorrell, C.C. (2011). "The effects of firing conditions on the properties of electrophoretically deposited titanium dioxide films on graphite substrates". Journal of the European Ceramic Society. 31 (15): 2877–2885. doi:10.1016/j.jeurceramsoc.2011.07.007

- Lapshin, R. V. (1998). "Automatic lateral calibration of tunneling microscope scanners" (PDF). Review of Scientific Instruments. 69 (9): 3268–3276. doi:10.1063/1.1149091

- "Weapons Lubricant in the Desert". September 16, 2005. Archived from the original on 2007-10-15. Retrieved 2009-06-06

Basics of Magnetic Ceramics

Magnetic ceramics possess strong magnetic coupling, low loss characteristics and high electrical resistivity. They are used in data storage, information transmission, power supply, etc. This chapter has been carefully written to provide an easy understanding of the varied facets of magnetic ceramics.

Magnetic Ceramics

Magnetic ceramics are important materials for a variety of applications such data storage, tunnel junctions, spin valves, high frequency applications etc. These materials possess extra-ordinary properties such as strong magnetic coupling, low loss characteristics and high electrical resistivity which is often related to their structure and composition. Depending upon the type of application, based on the knowledge of materials, one can choose appropriate material.

To build an understanding, we will first have a brief look at the basics of magnetic properties, followed by various types of magnetism in materials and key characteristics and differences. This will be followed by a discussion on magnetic ceramics with emphasis on some of the key applications.

Magnetic Moment

The magnetic moment of a magnet is a quantity that determines the torque it will experience in an external magnetic field. A loop of electric current, a bar magnet, an electron, a molecule, and a planet all have magnetic moments.

The magnetic moment may be considered to be a vector having a magnitude and direction. The direction of the magnetic moment points from the south to north pole of the magnet. The magnetic field produced by the magnet is proportional to its magnetic moment. More precisely, the term *magnetic moment* normally refers to a system's magnetic dipole moment, which produces the first term in the multipole expansion of a general magnetic field. The dipole component of an object's magnetic field is symmetric about the direction of its magnetic dipole moment, and decreases as the inverse cube of the distance from the object.

Definition

The magnetic moment is defined as a vector relating the aligning torque on the object from an externally applied magnetic field to the field vector itself. The relationship is given by:

$$\tau = \mu \times B$$

where τ is the torque acting on the dipole and B is the external magnetic field, and μ is the magnetic moment.

This definition is based on how one would measure the magnetic moment, in principle, of an unknown sample.

Units

The unit for magnetic moment is not a base unit in the International System of Units (SI). As the torque is measured in newton-meters (N·m) and the magnetic field in teslas (T), the magnetic moment is measured in newton-meters per tesla. This has equivalents in other base units:

$$N·m/T = A·m^2 = J/T$$

where A is amperes and J is joules.

In the CGS system, there are several different sets of electromagnetism units, of which the main ones are ESU, Gaussian, and EMU. Among these, there are two alternative (non-equivalent) units of magnetic dipole moment:

$$1 \; statA·cm^2 = 3.33564095 \times 10^{-14} \; A·m^2 \; (ESU)$$

$$1 \; erg/G = 1 \; abA·cm^2 = 10^{-3} \; A·m^2 \; (Gaussian \; and \; EMU),$$

where statA is statamperes, cm is centimeters, erg is ergs, G is gauss and abA is abamperes. The ratio of these two non-equivalent CGS units (EMU/ESU) is equal to the speed of light in free space, expressed in cm·s^{-1}.

All formulae in this article are correct in SI units; they may need to be changed for use in other unit systems. For example, in SI units, a loop of current with current I and area A has magnetic moment IA, but in Gaussian units the magnetic moment is IA/c.

Two Representations of the Cause of the Magnetic Moment

The preferred classical explanation of a magnetic moment has changed over time. Before the 1930s, textbooks explained the moment using hypothetical magnetic point charges. Since then, most have defined it in terms of Ampèrian currents. In magnetic materials, the cause of the magnetic moment are the spin and orbital angular momentum states of the electrons, and varies depending on whether atoms in one region are aligned with atoms in another.

Magnetic Pole Representation

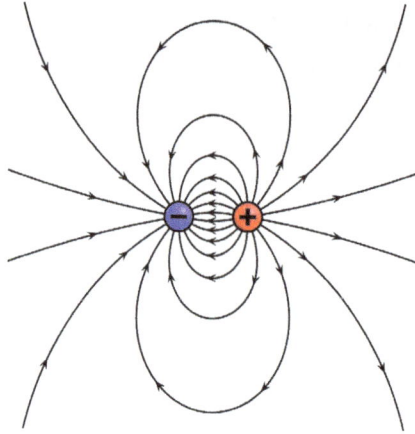

An electrostatic analog for a magnetic moment: two opposing charges separated by a finite distance.

The sources of magnetic moments in materials can be represented by poles in analogy to electrostatics. Consider a bar magnet which has magnetic poles of equal magnitude but opposite polarity. Each pole is the source of magnetic force which weakens with distance. Since magnetic poles always come in pairs, their forces partially cancel each other because while one pole pulls, the other repels. This cancellation is greatest when the poles are close to each other i.e. when the bar magnet is short. The magnetic force produced by a bar magnet, at a given point in space, therefore depends on two factors: the strength p of its poles (magnetic pole strength), and the vector l separating them. The moment is related to the fictitious poles as

$$\mu = p\mathbf{l}.$$

It points in the direction from South to North pole. The analogy with electric dipoles should not be taken too far because magnetic dipoles are associated with angular momentum. Nevertheless, magnetic poles are very useful for magnetostatic calculations, particularly in applications to ferromagnets. Practitioners using the magnetic pole approach generally represent the magnetic field by the irrotational field H, in analogy to the electric field E.

Integral Representation

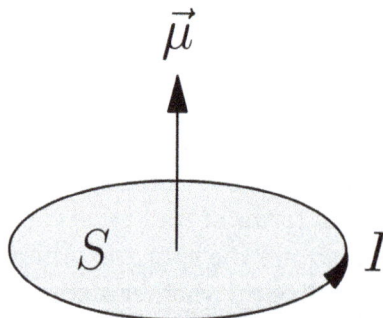

Moment **μ** of a planar current having magnitude **I** and enclosing an area **S**

We start from the definition of the differential magnetic moment pseudovector:

$$\mu = \tfrac{1}{2}\mathbf{r} \times \mathbf{j}$$

where $\times$ is the vector cross product, r is the position vector, and j is the electric current density. It is very similar to the differential angular momentum, defined as:

$$\mathbf{l} = \mathbf{r} \times (\rho\mathbf{v})$$

where ρ is the mass density and v is the velocity vector. Like in every pseudovector, by convention the direction of the cross product is given by the right hand grip rule. Practitioners using the current loop model generally represent the magnetic field by the solenoidal field B, analogous to the electrostatic field D.

The integral magnetic moment of a charge distribution is therefore:

$$M = \tfrac{1}{2}\iiint_v \mathbf{r} \times \mathbf{j}dV,$$

Let us start with a point particle; in this simple situation the magnetic moment is:

$$M = \tfrac{1}{2}q\mathbf{r} \times \mathbf{v},$$

wherer is the position of the electric chargeq relative to the center of the circle and v is the instantaneous velocity of the charge, giving an electric current density j.

On the other hand, for a point particle the angular momentum is defined as:

$$\mathbf{L} = \mathbf{r} \times \mathbf{p} = m\mathbf{r} \times \mathbf{v}_{,}$$

and in the planar case:

$$M = \tfrac{1}{2}\iint_s \mathbf{r} \times \mathbf{j}dS,$$

by defining the electric current with a vector areaS (the x-, y-, and z-coordinates of this vector are the areas of projections of the loop onto the yz-, zx-, and xy-planes):

$$M = \tfrac{1}{2}I\int_{\partial S} \mathbf{r} \times d\mathbf{r}.$$

Then by Stokes' theorem, integral magnetic moment then becomes expressible as:

$$M = I\mathbf{S}.$$

The factor 1/2 in our definition above is only due to historical reason: the old definition of the magnetic moment was this last integral equation. If one had started from a differential definition:

$$\mu = \mathbf{r} \times \mathbf{j}$$

then the coherent integral expression would have been:

$$M = 2I\mathbf{S}.$$

Magnetic Moment of a Solenoid

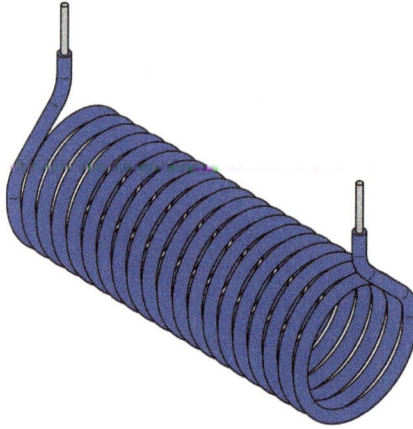

Image of a solenoid

A generalization of the above current loop is a coil, or solenoid. Its moment is the vector sum of the moments of individual turns. If the solenoid has N identical turns (single-layer winding) and vector area S,

$$\mu = N I \mathbf{S}.$$

Magnetic Moment and Angular Momentum

The magnetic moment has a close connection with angular momentum called the gyromagnetic effect. This effect is expressed on a macroscopic scale in the Einstein-de Haas effect, or "rotation by magnetization," and its inverse, the Barnett effect, or "magnetization by rotation." In particular, when a magnetic moment is subject to a torque in a magnetic field that tends to align it with the applied magnetic field, the moment precesses (rotates about the axis of the applied field). This is a consequence of the concomitance of magnetic moment and angular momentum, that in case of charged massive particles corresponds to the concomitance of charge and mass in a particle.

Viewing a magnetic dipole as a rotating charged particle brings out the close connection between magnetic moment and angular momentum. Both the magnetic moment and the angular momentum increase with the rate of rotation. The ratio of the two is called the gyromagnetic ratio and is simply the half of the charge-to-mass ratio.

For a spinning charged solid with a uniform charge density to mass density ratio, the gyromagnetic ratio is equal to half the charge-to-mass ratio. This implies that a more massive assembly of charges spinning with the same angular momentum will have a proportionately weaker magnetic moment, compared to its lighter counterpart. Even though atomic particles cannot be accurately described as spinning charge distributions of uniform charge-to-mass ratio, this general trend can be observed in the atomic world, where the intrinsic angular momentum (spin) of each type of particle is a constant: a small half-integer times the reduced Planck constant $\hbar$. This is the basis for defining the magnetic moment units of Bohr magneton (assuming charge-to-mass ratio of the electron) and nuclear magneton (assuming charge-to-mass ratio of the proton).

Effects of an External Magnetic Field on a Magnetic Moment

Force on a Moment

A magnetic moment in an externally produced magnetic field has a potential energy U:

$$U = -\mu \cdot B$$

In a case when the external magnetic field is non-uniform, there will be a force, proportional to the magnetic field gradient, acting on the magnetic moment itself. There has been some discussion on how to calculate the force acting on a magnetic dipole. There are two expressions for the force acting on a magnetic dipole, depending on whether the model used for the dipole is a current loop or two monopoles (analogous to the electric dipole). The force obtained in the case of a current loop model is

$$\mathbf{F}_{loop} = \nabla(\mu \cdot \mathbf{B})$$

In the case of a pair of monopoles being used (i.e. electric dipole model)

$$\mathbf{F}_{dipole} = (\mu \cdot \nabla)\mathbf{B}$$

and one can be put in terms of the other via the relation

$$\mathbf{F}_{loop} = \mathbf{F}_{dipole} + \mu \times (\nabla \times \mathbf{B})$$

In all these expressions μ is the dipole and B is the magnetic field at its position. Note that if there are no currents or time-varying electrical fields $\nabla \times B = 0$ and the two expressions agree.

An electron, nucleus, or atom placed in a uniform magnetic field will precess with a frequency known as the Larmor frequency.

Magnetic Dipoles

A magnetic dipole is the limit of either a current loop or a pair of poles as the dimensions of the source are reduced to zero while keeping the moment constant. As long as these limits only apply to fields far from the sources, they are equivalent. However, the two models give different predictions for the internal field.

External Magnetic Field Produced by a Magnetic Dipole Moment

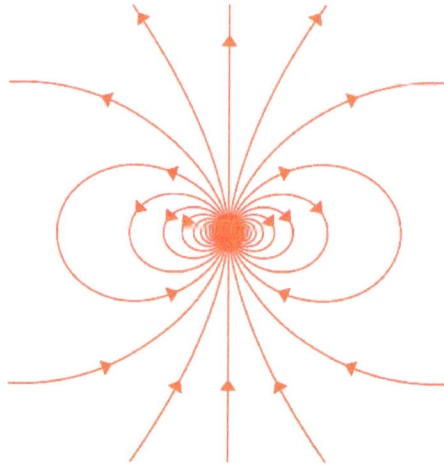

Magnetic field lines around a "magnetostatic dipole". The magnetic dipole itself is located in the center of the figure, seen from the side, and pointing upward.

Any system possessing a net magnetic dipole moment m will produce a dipolar magnetic field (described below) in the space surrounding the system. While the net magnetic field produced by the system can also have higher-order multipole components, those will drop off with distance more rapidly, so that only the dipolar component will dominate the magnetic field of the system at distances far away from it.

The vector potential of magnetic field produced by magnetic moment m is

$$\mathbf{A}(\mathbf{r}) = \frac{\mu_0}{4\pi} \frac{\mathbf{m} \times \mathbf{r}}{|\mathbf{r}|^3}$$

and magnetic flux density is

$$\mathbf{B}(\mathbf{r}) = \nabla \times \mathbf{A} = \frac{\mu_0}{4\pi} \left(\frac{3\mathbf{r}(\mathbf{m} \cdot \mathbf{r})}{|\mathbf{r}|^5} - \frac{\mathbf{m}}{|\mathbf{r}|^3} \right).$$

Alternatively one can obtain the scalar potential first from the magnetic pole perspective,

$$\psi(\mathbf{r}) = \frac{\mathbf{m} \cdot \mathbf{r}}{4\pi |\mathbf{r}|^3},$$

and hence magnetic field strength is

$$\mathbf{H}(\mathbf{r}) = -\nabla \psi = \frac{1}{4\pi} \left(\frac{3\mathbf{r}(\mathbf{m} \cdot \mathbf{r})}{|\mathbf{r}|^5} - \frac{\mathbf{m}}{|\mathbf{r}|^3} \right).$$

The magnetic field of an ideal magnetic dipole is depicted on the right.

Internal Magnetic Field of a Dipole

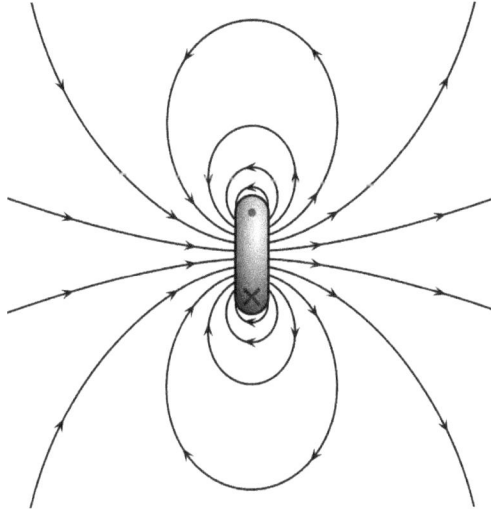

The magnetic field of a current loop

The two models for a dipole (current loop and magnetic poles) give the same predictions for the magnetic field far from the source. However, inside the source region they give different predictions. The magnetic field between poles is in the opposite direction to the magnetic moment (which points from the negative charge to the positive charge), while inside a current loop it is in the same direction. Clearly, the limits of these fields must also be different as the sources shrink to zero size. This distinction only matters if the dipole limit is used to calculate fields inside a magnetic material.

If a magnetic dipole is formed by making a current loop smaller and smaller, but keeping the product of current and area constant, the limiting field is

$$\mathbf{B}(\mathbf{x}) = \frac{\mu_0}{4\pi} \left[\frac{3\mathbf{n}(\mathbf{n} \cdot \mathbf{m}) - \mathbf{m}}{|\mathbf{x}|^3} + \frac{8\pi}{3} \mathbf{m}\delta(\mathbf{x}) \right].$$

This limit is correct for the internal field of the dipole.

If a magnetic dipole is formed by taking a "north pole" and a "south pole", bringing them closer and closer together but keeping the product of magnetic pole-charge and distance constant, the limiting field is

$$\mathbf{H}(\mathbf{x}) = \frac{1}{4\pi} \left[\frac{3\mathbf{n}(\mathbf{n} \cdot \mathbf{m}) - \mathbf{m}}{|\mathbf{x}|^3} - \frac{4\pi}{3} \mathbf{m}\delta(\mathbf{x}) \right].$$

These fields are related by $\mathbf{B} = \mu_0(\mathbf{H} + \mathbf{M})$, where $\mathbf{M}(\mathbf{x}) = \mathbf{m}\delta(\mathbf{x})$ is the magnetization.

Forces between Two Magnetic Dipoles

As discussed earlier, the force exerted by a dipole loop with moment $\mathbf{m}_1$ on another with moment $\mathbf{m}_2$ is

$$\mathbf{F} = \nabla(\mathbf{m}_2 \cdot \mathbf{B}_1),$$

where $\mathbf{B}_1$ is the magnetic field due to moment $\mathbf{m}_1$. The result of calculating the gradient is

$$\mathbf{F}(\mathbf{r}, \mathbf{m}_1, \mathbf{m}_2) = \frac{3\mu_0}{4\pi |\mathbf{r}|^4} \left(\mathbf{m}_2(\mathbf{m}_1 \cdot \hat{\mathbf{r}}) + \mathbf{m}_1(\mathbf{m}_2 \cdot \hat{\mathbf{r}}) + \hat{\mathbf{r}}(\mathbf{m}_1 \cdot \mathbf{m}_2) - 5\hat{\mathbf{r}}(\mathbf{m}_1 \cdot \hat{\mathbf{r}})(\mathbf{m}_2 \cdot \hat{\mathbf{r}}) \right),$$

where $\hat{\mathbf{r}}$ is the unit vector pointing from magnet 1 to magnet 2 and r is the distance. An equivalent expression is

$$\mathbf{F} = \frac{3\mu_0}{4\pi |\mathbf{r}|^4} \left((\hat{\mathbf{r}} \times \mathbf{m}_1) \times \mathbf{m}_2 + (\hat{\mathbf{r}} \times \mathbf{m}_2) \times \mathbf{m}_1 - 2\hat{\mathbf{r}}(\mathbf{m}_1 \cdot \mathbf{m}_2) + 5\hat{\mathbf{r}}(\hat{\mathbf{r}} \times \mathbf{m}_1) \cdot (\hat{\mathbf{r}} \times \mathbf{m}_2) \right).$$

The force acting on $\mathbf{m}_1$ is in the opposite direction.

The torque of magnet 1 on magnet 2 is

$$\tau = \mathbf{m}_2 \times \mathbf{B}_1$$

Examples of Magnetic Moments

Two Kinds of Magnetic Sources

Fundamentally, contributions to any system's magnetic moment may come from sources of two kinds: motion of electric charges, such as electric currents; and the intrinsic magnetism of elementary particles, such as the electron.

Contributions due to the sources of the first kind can be calculated from knowing the distribution of all the electric currents (or, alternatively, of all the electric charges and their velocities) inside the system, by using the formulas below. On the other hand, the magnitude of each elementary particle's intrinsic magnetic moment is a fixed number, often measured experimentally to a great precision. For example, any electron's magnetic moment is measured to be $-9.284764 \times 10^{-24}$ J/T. The direction of the magnetic moment of any elementary particle is entirely determined by the direction of its spin, with the negative value indicating that any electron's magnetic moment is antiparallel to its spin.

The net magnetic moment of any system is a vector sum of contributions from one or both types of sources. For example, the magnetic moment of an atom of hydrogen-1 (the lightest hydrogen isotope, consisting of a proton and an electron) is a vector sum of the following contributions:

1. the intrinsic moment of the electron,

2. the orbital motion of the electron around the proton,

3. the intrinsic moment of the proton.

Similarly, the magnetic moment of a bar magnet is the sum of the contributing magnetic moments, which include the intrinsic and orbital magnetic moments of the unpaired electrons of the magnet's material and the nuclear magnetic moments.

Magnetic Moment of an atom

For an atom, individual electron spins are added to get a total spin, and individual orbital angular momenta are added to get a total orbital angular momentum. These two then are added using angular momentum coupling to get a total angular momentum. For an atom with no nuclear magnetic moment, the magnitude of the atomic dipole moment is then

$$m_{Atom} = g_J \mu_B \sqrt{j(j+1)}$$

where j is the total angular momentum quantum number, g_J is the Landé g-factor, and μ_B is the Bohr magneton. The component of this magnetic moment along the direction of the magnetic field is then

$$m_{Atom}(z) = -m g_J \mu_B$$

where m is called the magnetic quantum number or the *equatorial* quantum number, which can take on any of $2j + 1$ values:

$$-j, -(j-1) \cdots 0 \cdots + (j-1), +j.$$

The negative sign occurs because electrons have negative charge.

Due to the angular momentum, the dynamics of a magnetic dipole in a magnetic field differs from that of an electric dipole in an electric field. The field does exert a torque on the magnetic dipole tending to align it with the field. However, torque is proportional to rate of change of angular momentum, so precession occurs: the direction of spin changes. This behavior is described by the Landau–Lifshitz–Gilbert equation:

$$\frac{1}{\gamma}\frac{d\mathbf{m}}{dt} = \mathbf{m} \times \mathbf{H}_{\mathit{eff}} - \frac{\lambda}{\gamma m}\mathbf{m} \times \frac{d\mathbf{m}}{dt}$$

where γ is the gyromagnetic ratio, m is the magnetic moment, λ is the damping coefficient and $\mathbf{H}_{\mathit{eff}}$ is the effective magnetic field (the external field plus any self-induced field). The first term describes precession of the moment about the effective field, while the second is a damping term related to dissipation of energy caused by interaction with the surroundings.

Magnetic Moment of an Electron

Electrons and many elementary particles also have intrinsic magnetic moments, an explanation of which requires a quantum mechanical treatment and relates to the intrinsic angular momentum of the particles as discussed in the article Electron magnetic moment. It is these intrinsic magnetic moments that give rise to the macroscopic effects of magnetism, and other phenomena, such as electron paramagnetic resonance.

The magnetic moment of the electron is

$$\mathbf{m}_S = -\frac{g_S \mu_B \mathbf{S}}{\hbar},$$

where μ_B is the Bohr magneton, S is electron spin, and the g-factor g_S is 2 according to Dirac's theory, but due to quantum electrodynamic effects it is slightly larger in reality: 2.00231930436. The deviation from 2 is known as the anomalous magnetic dipole moment.

Again it is important to notice that m is a negative constant multiplied by the spin, so the magnetic moment of the electron is antiparallel to the spin. This can be understood with the following classical picture: if we imagine that the spin angular momentum is created by the electron mass spinning around some axis, the electric current that this rotation creates circulates in the opposite direction, because of the negative charge of the electron; such current loops produce a magnetic moment which is antiparallel to the spin. Hence, for a positron (the anti-particle of the electron) the magnetic moment is parallel to its spin.

Magnetic Moment of a Nucleus

The nuclear system is a complex physical system consisting of nucleons, i.e., protons and neutrons. The quantum mechanical properties of the nucleons include the spin among others. Since the electromagnetic moments of the nucleus depend on the spin of the individual nucleons, one can look at these properties with measurements of nuclear moments, and more specifically the nuclear magnetic dipole moment.

Most common nuclei exist in their ground state, although nuclei of some isotopes have long-lived excited states. Each energy state of a nucleus of a given isotope is characterized by a well-defined magnetic dipole moment, the magnitude of which is a fixed number, often measured experimentally to a great precision. This number is very sensitive to the individual contributions from nucleons, and a measurement or prediction of its value can reveal important information about the content of the nuclear wave function. There are several theoretical models that predict the value of the magnetic dipole moment and a number of experimental techniques aiming to carry out measurements in nuclei along the nuclear chart.

Magnetic Moment of a Molecule

Any molecule has a well-defined magnitude of magnetic moment, which may depend on the molecule's energy state. Typically, the overall magnetic moment of a molecule is a combination of the following contributions, in the order of their typical strength:

- magnetic moments due to its unpaired electron spins (paramagnetic contribution), if any

- orbital motion of its electrons, which in the ground state is often proportional to the external magnetic field (diamagnetic contribution)

- the combined magnetic moment of its nuclear spins, which depends on the nuclear spin configuration.

Examples of Molecular Magnetism

- The dioxygen molecule, O_2, exhibits strong paramagnetism, due to unpaired spins of its outermost two electrons.

- The carbon dioxide molecule, CO_2, mostly exhibits diamagnetism, a much weaker magnetic moment of the electron orbitals that is proportional to the external magnetic field. The nuclear magnetism of a magnetic isotope such as ^{13}C or ^{17}O will contribute to the molecule's magnetic moment.

- The dihydrogen molecule, H_2, in a weak (or zero) magnetic field exhibits nuclear magnetism, and can be in a para- or an ortho- nuclear spin configuration.

- Many transition metal complexes are magnetic. The spin-only formula is a good first approximation for high-spin complexes of first-row transition metals.

Number of unpaired electrons	Spin-only moment (μ_B)
1	1.73
2	2.83
3	3.87
4	4.90
5	5.92

Elementary Particles

In atomic and nuclear physics, the Greek symbol μ represents the magnitude of the magnetic moment, often measured in Bohr magnetons or nuclear magnetons, associated with the intrinsic spin of the particle and/or with the orbital motion of the particle in a system. Values of the intrinsic magnetic moments of some particles are given in the table below:

Intrinsic magnetic moments and spins of some elementary particles		
Particle	Magnetic dipole moment (10^{-27} J·T^{-1})	Spin quantum number (dimensionless)
electron (e^-)	−9284.764	1/2
proton (H^+)	14.106067	1/2
neutron (n)	−9.66236	1/2
muon (μ^-)	−44.904478	1/2
deuteron ($^2H^+$)	4.3307346	1
triton ($^3H^+$)	15.046094	1/2
helion ($^3He^{2+}$)	−10.746174	1/2
alpha particle ($^4He^{2+}$)	0	0

Magnetism in materials is crudely explained as mutual attraction between two pieces of a material, say iron or iron ore. There are various microscopic mechanisms of magnetism in materials. The strength of magnetism is quantitatively judged by a quantity called as 'magnetic moment'.

The major contributors of magnetic moment in a material are

- Motion of electrons in an orbit of an atom. Orbital moment can be related to the current flowing in a loop of a wire of zero (negligible) resistance.

- Spinning of its electron around it own spin axis gives rise to a moment.

- Nuclear magnetic moment due to nuclei.

The first two contributions are quite significant and contribute to most of the magnetic character of a material while the third component, nuclear magnetic moment, is rather insignificant in the context of most magnetic materials of practical interest and can be neglected.

Magnetization

In classical electromagnetism, magnetization or magnetic polarization is the vector field that expresses the density of permanent or induced magnetic dipole moments in a magnetic material. The origin of the magnetic moments responsible for magnetization can be either microscopic electric currents resulting from the motion of electrons in atoms, or the spin of the electrons or the nuclei. Net magnetization results from the response of a material to an external magnetic field, together with any unbalanced magnetic dipole moments that may be inherent in the material itself; for example, in ferromagnets. Magnetization is not always uniform within a body, but rather varies between different points. Magnetization also describes how a material responds to an applied magnetic field as well as the way the material changes the magnetic field, and can be used to calculate the forces that result from those interactions. It can be compared to electric polarization, which is the measure of the corresponding response of a material to an electric field in electrostatics. Physicists and engineers usually define magnetization as the quantity of magnetic moment per unit volume. It is represented by a pseudovector M.

Definition

The magnetization field or M-field can be defined according to the following equation:

$$\mathbf{M} = \frac{d\mathbf{m}}{dV}$$

Where $d\mathbf{m}$ is the elementary magnetic moment and dV is the volume element; in other words, the M-field is the distribution of magnetic moments in the region or manifold concerned. This is better illustrated through the following relation:

$$\mathbf{m} = \iiint \mathbf{M} dV$$

where m is an ordinary magnetic moment and the triple integral denotes integration over a volume. This makes the M-field completely analogous to the electric polarisation field, or P-field, used to determine the electric dipole moment p generated by a similar region or manifold with such a polarization:

$$\mathbf{P} = \frac{d\mathbf{p}}{dV}, \quad \mathbf{p} = \iiint \mathbf{P} dV$$

Where $d\mathbf{p}$ is the elementary electric dipole moment.

Those definitions of P and M as a "moments per unit volume" are widely adopted, though in some cases they can lead to ambiguities and paradoxes.

The M-field is measured in *amperes per meter* (A/m) in SI units.

Physics Application

The magnetization is often not listed as a material parameter for commercially available ferromagnets. Instead the parameter that is listed is residual flux density, denoted $\mathbf{B_r}$. Physicists often need the magnetization to calculate the moment of a ferromagnet. To calculate the dipole moment m (A m²) using the formula:

$$\mathbf{m} = \mathbf{M}V,$$

we have that

$$\mathbf{M} = \mathbf{B_r} / \mu_0,$$

thus

$$\mathbf{m} = \mathbf{B_r}V / \mu_0,$$

where:

- is the Residual Flux Density, expressed in Teslas (T).

- is the volume (m³) of the magnet.

- H/m is the permeability of vacuum.

Magnetization in Maxwell's Equations

The behavior of magnetic fields (B, H), electric fields (E, D), charge density (ρ), and current density (J) is described by Maxwell's equations. The role of the magnetization is described below.

Relations between B, H, and M

The magnetization defines the auxiliary magnetic field H as

$$\mathbf{B} = \mu_0(\mathbf{H} + \mathbf{M}) \text{ (SI units)}$$

$$\mathbf{B} = \mathbf{H} + 4\pi\mathbf{M} \text{ (Gaussian units)}$$

which is convenient for various calculations. The vacuum permeability μ_0 is, by definition, $4\pi \times 10^{-7}$ V·s/(A·m).

A relation between M and H exists in many materials. In diamagnets and paramagnets, the relation is usually linear:

$$\mathbf{M} = \chi_m \mathbf{H}$$

where χ_m is called the volume magnetic susceptibility.

In ferromagnets there is no one-to-one correspondence between M and H because of Magnetic hysteresis.

Magnetization Current

The magnetization M makes a contribution to the current density J, known as the magnetization current.

$$\mathbf{J} = \nabla \times \mathbf{M}$$

and for the bound surface current:

$$\mathbf{K_m} = \mathbf{M} \times \hat{\mathbf{n}}$$

so that the total current density that enters Maxwell's equations is given by

$$\mathbf{J} = \mathbf{J_f} + \nabla \times \mathbf{M} + \frac{\partial \mathbf{P}}{\partial t}$$

where J_f is the electric current density of free charges (also called the free current), the second term is the contribution from the magnetization, and the last term is related to the electric polarization P.

Magnetostatics

In the absence of free electric currents and time-dependent effects, Maxwell's equations describing the magnetic quantities reduce to

$$\nabla \times \mathbf{H} = 0$$
$$\nabla \cdot \mathbf{H} = -\nabla \cdot \mathbf{M}$$

These equations can be solved in analogy with electrostatic problems where

$$\nabla \cdot \mathbf{E} = \frac{\rho}{\epsilon_0}$$
$$\nabla \times \mathbf{E} = 0$$

In this sense $-\nabla \cdot M$ plays the role of a fictitious "magnetic charge density" analogous to the electric charge density ρ.

It is important to note that there is no such thing as a "magnetic charge," but that issue was still debated through the whole 19th century. Other concepts, that went along with it, such as the auxiliary field H, also have no real physical meaning in their own right. However, they are convenient mathematical tools, and are therefore still used today for applications such as modeling the magnetic field of the Earth.

Magnetization Dynamics

The time-dependent behavior of magnetization becomes important when considering nanoscale and nanosecond timescale magnetization. Rather than simply aligning with an applied field, the individual magnetic moments in a material begin to precess around the applied field and come into alignment through relaxation as energy is transferred into the lattice.

Reversal

Magnetization reversal, also known as switching, refers to the process that leads to a 180° (arc) re-orientation of the magnetization vector with respect to its initial direction, from one stable orientation to the opposite one. Technologically, this is one of the most important processes in magnetism that is linked to the magnetic data storage process such as used in modern hard disk drives. As it is known today, there are only a few possible ways to reverse the magnetization of a metallic magnet:

1. an applied magnetic field

2. spin injection via a beam of particles with spin

3. magnetization reversal by circularly polarized light; i.e., incident electromagnetic radiation that is circularly polarized

Demagnetization

Demagnetization is the reduction or elimination of magnetization. One way to do this is to heat the object above its Curie temperature, where thermal fluctuations have enough energy to overcome exchange interactions, the source of ferromagnetic order, and destroy that order. Another way is to pull it out of an electric coil with alternating current running through it, giving rise to fields that oppose the magnetization.

One application of demagnetization is to eliminate unwanted magnetic fields. For example, magnetic fields can interfere with electronic devices such as cell phones or computers, and with machining by making cuttings cling to their parent.

Macroscopic View of Magnetization

We define magnetic induction, B, as $\mu_0 H$ where μ_0 is magnetic permeability of vacuum and its value is equal to $4\pi * 10^{-7} H.m^{-1}$. Units of B are tesla or Weber.m^{-2}.

This also explains that while H depends only on the current, B also depends upon the medium surrounding the wire which defines μ_0.

So now rewriting equation yields

$$B = \mu_\circ NI$$

Now imagine if a magnetic material is inserted within the coil, then there is a current induced in the magnetic material too, called Ameprian current, I_a which modifies the above equation to

$$B = \mu_\circ NI + \mu_\circ I_a$$

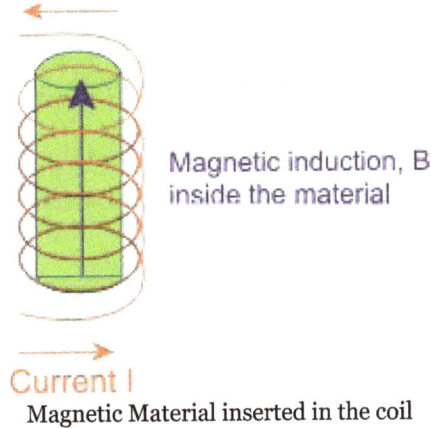

Magnetic induction, B inside the material

Current I

Magnetic Material inserted in the coil

This Amperain current in the magnetic material can be replaced with induced magnetization, M, and hence using the above equation, we can write

$$B = \mu_\circ (H + M)$$

If magnetization is assumed to be proportional to the magnetizing field, H, with proportionality constant defined as magnetic susceptibility, χ_m i.e. $M = \chi_m . H$, we can write

$$B = \mu_\circ (H + \chi_m M)$$

OR

$$B = \mu_\circ H (1 + \chi_m)$$

OR

$$B = \mu_\circ \mu_r H$$

Here, χ_m, similar to dielectric materials, can be thought of as a parameter which expresses magnetic response of electron in a material to the applied magnetic field and is a dimensionless quantity. Here, $\mu_r = (1 + \chi_m)$ is the, in a similar manner to relative dielectric permittivity, ε_r.

In general, both susceptibility and permeability are tensors and assuming that vectors are collinear wherever there is vector notation is not used.

Naturally for vacuum, $\chi_m = 0$.. However, unlike dielectric materials, χ_m can acquire both positive and negative values.

Classification of Magnetism

Magnetic materials can be classified based on the values of magnetic susceptibility.

Materials with negative susceptibility are diamagnetic ($\chi_m < 1$). Most diamagnetic materials show very small negative susceptibility except superconductors in superconducting effect when χ_m is equal to 1 which is very useful for applications such as magnetic levitation.

Materials with positive susceptibility are either paramagnetic, ferromagnetic or ferris-magnetic ($\chi_m > 1$). Susceptibility is positive but very small for paramagnetic materials but can be very large for ferro- and ferri-magnetic materials.

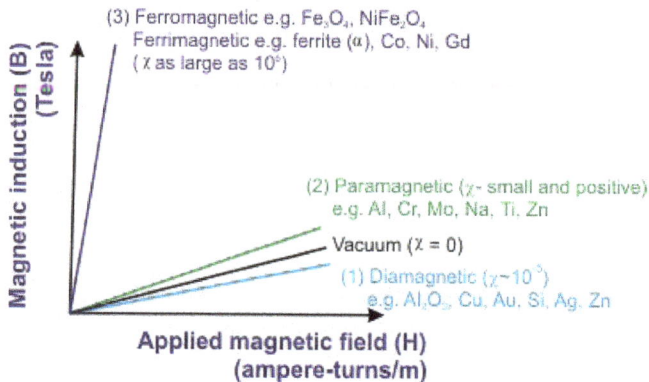

Classification of materials on the basis of susceptibility

Susceptibility values of some of the common materials are provided below.

Material	χ(SI) unitless	χ(cgs) Unitless	μ Unitless	Type of Magnetism
Bi	-165×10^{-6}	-13.13×10^{-6}	0.99983	
Be	-23.2×10^{-6}	-1.85×10^{-6}	0.99998	
Ag	-23.2×10^{-6}	-1.90×10^{-6}	0.99997	
Au	-34.4×10^{-6}	-2.74×10^{-6}	0.99996	
Ge	-34.4×10^{-6}	-5.66×10^{-6}	0.99999	Diamagnetic
Cu	-9.7×10^{-6}	-0.77×10^{-6}	0.99999	
Si	-4.1×10^{-6}	-0.32×10^{-6}	0.99999	
Water	-9.14×10^{-6}	-0.73×10^{-6}	0.99999	
Superconductors (only in superconducting state)	-1.0	$\sim -8 \times 10^{-2}$	0	

β-Sn	$+2.4 \times 10^{-6}$	$+0.19 \times 10^{-6}$	1	
W	$+77.7 \times 10^{-6}$	$+6.18 \times 10^{-6}$	1.00008	**Paramagnetic**
Al	$+20.7 \times 10^{-6}$	$+1.65 \times 10^{-6}$	1.00002	
Pt	$+264.4 \times 10^{-6}$	$+21.04 \times 10^{-6}$	1.000026	
Low carbon steel	$\approx 5 \times 10^3$	3.98×10^2	5×10^3	
Fe-3%Si (Grain Oriented)	4×10^3	3.18×10^3	4×10^4	**Ferromagnetic**
Ni-Fe-Mo superalloy	10^6	7.96×10^4	10^6	

It should be noted, also as we will see, that except for diamagnetic materials, magnetic susceptibilities are temperature dependent. Sign of susceptibility can also be related (qualitatively) to the penetration of magnetic flux inside the material.

This says that for diamagnetic materials, when an external field is applied, the magnetic moment that is induced is in opposite direction to the field direction i.e. opposite magnetization as shown below. This is an inherent effect present in all materials. It is just that some materials like silver, gold, silicon are only diamagnetic i.e. they don't have any other effect present in them.

In many other materials, on top of the diamagnetic effect which is inherent to all materials, other effects are present which contribute significantly to the magnetization and all of these tend to have induced magnetization that is in the direction of the applied i.e. positive susceptibility. This means that magnetic flux penetrates into the material as shown below. We will discuss these effects one by one.

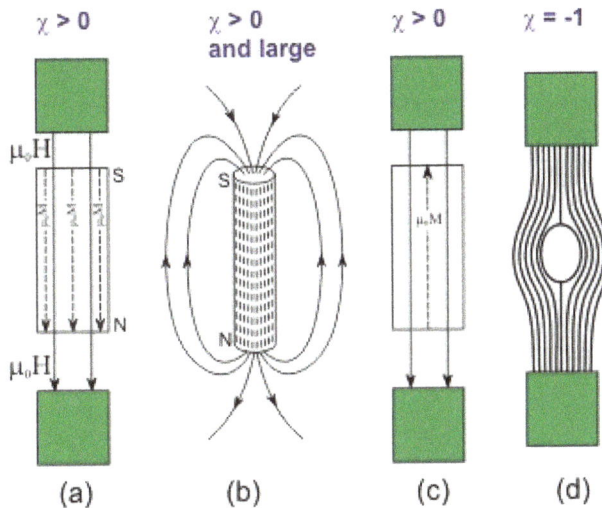

Schematic diagram showing flux penetration in magnetic materials with various ranges of susceptibilities.

In a nutshell, the net susceptibility of a material is the sum of all types of magnetic contributions. For materials having only diamagnetic contribution, this turns out to be negative.

Moreover, it is quantum mechanics which shows that materials are diamagnetic or paramagnetic or ferromagnetic. If taken too far, classical mechanics simply shows that all the magnetic moments in any material cancel out.

Diamagnetism

Diamagnetic materials are repelled by a magnetic field; an applied magnetic field creates an induced magnetic field in them in the opposite direction, causing a repulsive force. In contrast, paramagnetic and ferromagnetic materials are attracted by a magnetic field. Diamagnetism is a quantum mechanical effect that occurs in all materials; when it is the only contribution to the magnetism the material is called diamagnetic. In paramagnetic and ferromagnetic substances the weak diamagnetic force is overcome by the attractive force of magnetic dipoles in the material. The magnetic permeability of diamagnetic materials is less than μ_o, the permeability of vacuum. In most materials diamagnetism is a weak effect which can only be detected by sensitive laboratory instruments, but a superconductor acts as a strong diamagnet because it repels a magnetic field entirely from its interior.

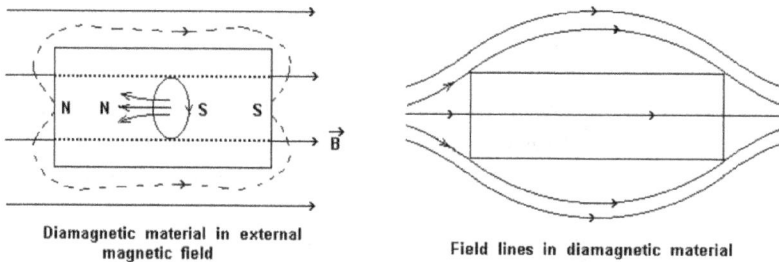

Diamagnetic material in external
magnetic field

Field lines in diamagnetic material

Dimagnetic material interaction in magnetic field.

Diamagnetism was first discovered when Sebald Justinus Brugmans observed in 1778 that bismuth and antimony were repelled by magnetic fields. In 1845, Michael Faraday demonstrated that it was a property of matter and concluded that every material responded (in either a diamagnetic or paramagnetic way) to an applied magnetic field. He adopted the term *diamagnetism* after it was suggested to him by William Whewell.

Materials

Notable diamagnetic materials	
Material	χ_v **($\times$ 10^{-5})**
Superconductor	-10^5
Pyrolytic carbon	-40.9
Bismuth	-16.6
Mercury	-2.9

Notable diamagnetic materials	
Material	χ_v ($\times 10^{-5}$)
Silver	−2.6
Carbon (diamond)	−2.1
Lead	−1.8
Carbon (graphite)	−1.6
Copper	−1.0
Water	−0.91

Diamagnetism, to a greater or lesser degree, is a property of all materials and always makes a weak contribution to the material's response to a magnetic field. For materials that show some other form of magnetism (such as ferromagnetism or paramagnetism), the diamagnetic contribution becomes negligible. Substances that mostly display diamagnetic behaviour are termed diamagnetic materials, or diamagnets. Materials called diamagnetic are those that laymen generally think of as *non-magnetic*, and include water, wood, most organic compounds such as petroleum and some plastics, and many metals including copper, particularly the heavy ones with many core electrons, such as mercury, gold and bismuth. The magnetic susceptibility values of various molecular fragments are called Pascal's constants.

Diamagnetic materials, like water, or water-based materials, have a relative magnetic permeability that is less than or equal to 1, and therefore a magnetic susceptibility less than or equal to 0, since susceptibility is defined as $\chi_v = \mu_v - 1$. This means that diamagnetic materials are repelled by magnetic fields. However, since diamagnetism is such a weak property, its effects are not observable in everyday life. For example, the magnetic susceptibility of diamagnets such as water is $\chi_v = -9.05 \times 10^{-6}$. The most strongly diamagnetic material is bismuth, $\chi_v = -1.66 \times 10^{-4}$, although pyrolytic carbon may have a susceptibility of $\chi_v = -4.00 \times 10^{-4}$ in one plane. Nevertheless, these values are orders of magnitude smaller than the magnetism exhibited by paramagnets and ferromagnets. Note that because χ_v is derived from the ratio of the internal magnetic field to the applied field, it is a dimensionless value.

All conductors exhibit an effective diamagnetism when they experience a changing magnetic field. The Lorentz force on electrons causes them to circulate around forming eddy currents. The eddy currents then produce an induced magnetic field opposite the applied field, resisting the conductor's motion.

Superconductors

Superconductors may be considered perfect diamagnets ($\chi_v = -1$), because they expel all fields (except in a thin surface layer) due to the Meissner effect.

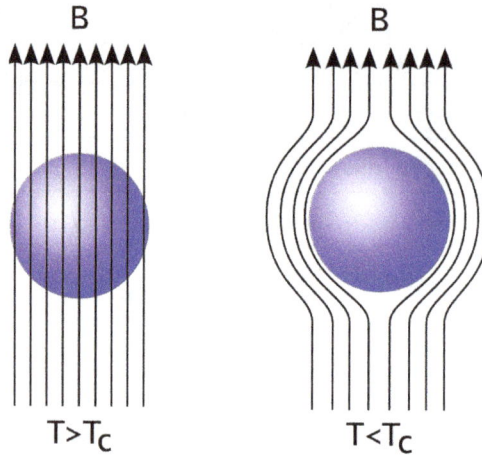

Transition from ordinary conductivity (left) to superconductivity (right). At the transition, the superconductor expels the magnetic field and then acts as a perfect diamagnet.

Demonstrations

Curving Water Surfaces

If a powerful magnet (such as a supermagnet) is covered with a layer of water (that is
thin compared to the diameter of the magnet) then the field of the magnet significantly
repels the water. This causes a slight dimple in the water's surface that may be seen by
its reflection.

Levitation

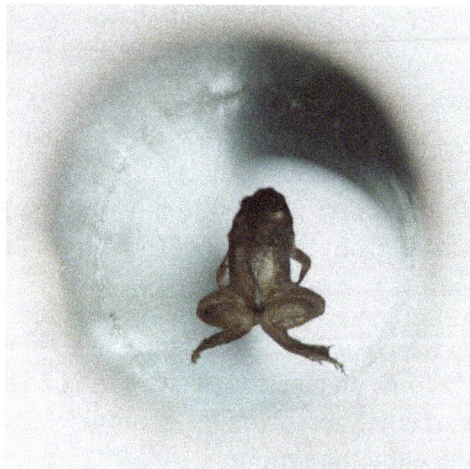

A live frog levitates inside a 32 mm (1.26 in) diameter vertical bore of a Bitter solenoid in a magnetic field
of about 16 teslas at the Nijmegen High Field Magnet Laboratory.

Diamagnets may be levitated in stable equilibrium in a magnetic field, with no power consumption. Earnshaw's theorem seems to preclude the possibility of static magnetic levitation. However, Earnshaw's theorem applies only to objects with positive

susceptibilities, such as ferromagnets (which have a permanent positive moment) and paramagnets (which induce a positive moment). These are attracted to field maxima, which do not exist in free space. Diamagnets (which induce a negative moment) are attracted to field minima, and there can be a field minimum in free space.

A thin slice of pyrolytic graphite, which is an unusually strong diamagnetic material, can be stably floated in a magnetic field, such as that from rare earth permanent magnets. This can be done with all components at room temperature, making a visually effective demonstration of diamagnetism.

The Radboud University Nijmegen, the Netherlands, has conducted experiments where water and other substances were successfully levitated. Most spectacularly, a live frog was levitated.

In September 2009, NASA's Jet Propulsion Laboratory in Pasadena, California announced it had successfully levitated mice using a superconducting magnet, an important step forward since mice are closer biologically to humans than frogs. JPL said it hopes to perform experiments regarding the effects of microgravity on bone and muscle mass.

Recent experiments studying the growth of protein crystals have led to a technique using powerful magnets to allow growth in ways that counteract Earth's gravity.

A simple homemade device for demonstration can be constructed out of bismuth plates and a few permanent magnets that levitate a permanent magnet.

Theory

The electrons in a material generally circulate in orbitals, with effectively zero resistance and act like current loops. Thus it might be imagined that diamagnetism effects in general would be very, very common, since any applied magnetic field would generate currents in these loops that would oppose the change, in a similar way to superconductors, which are essentially perfect diamagnets. However, since the electrons are rigidly held in orbitals by the charge of the protons and are further constrained by the Pauli exclusion principle, many materials exhibit diamagnetism, but typically respond very little to the applied field.

The Bohr–van Leeuwen theorem proves that there cannot be any diamagnetism or paramagnetism in a purely classical system. However, the classical theory for Langevin diamagnetism gives the same prediction as the quantum theory. The classical theory is given below.

Langevin Diamagnetism

The Langevin theory of diamagnetism applies to materials containing atoms with closed shells. A field with intensity B, applied to an electron with charge e and mass

m, gives rise to Larmor precession with frequency $\omega = eB/2m$. The number of revolutions per unit time is $\omega/2\pi$, so the current for an atom with Z electrons is (in SI units)

$$I = -\frac{Ze^2 B}{4\pi m}.$$

The magnetic moment of a current loop is equal to the current times the area of the loop. Suppose the field is aligned with the z axis. The average loop area can be given as $\pi\langle\rho^2\rangle$, where $\langle\rho^2\rangle$ is the mean square distance of the electrons perpendicular to the z axis. The magnetic moment is therefore

$$\mu = -\frac{Ze^2 B}{4m}\langle\rho^2\rangle.$$

If the distribution of charge is spherically symmetric, we can suppose that the distribution of x,y,z coordinates are independent and identically distributed. Then $\langle x^2\rangle = \langle y^2\rangle = \langle z^2\rangle = \frac{1}{3}\langle r^2\rangle$, where $\langle r^2\rangle$ is the mean square distance of the electrons from the nucleus. Therefore, $\langle\rho^2\rangle = \langle x^2\rangle + \langle y^2\rangle = \frac{2}{3}\langle r^2\rangle$. If N is the number of atoms per unit volume, the diamagnetic susceptibility in SI units is

$$\chi = \frac{\mu_0 N\mu}{B} = -\frac{\mu_0 NZe^2}{6m}\langle r^2\rangle.$$

In Metals

The Langevin theory does not apply to metals because they have non-localized electrons. The theory for the diamagnetism of a free electron gas is called Landau diamagnetism, and instead considers the weak counter-acting field that forms when their trajectories are curved due to the Lorentz force. Landau diamagnetism, however, should be contrasted with Pauli paramagnetism, an effect associated with the polarization of delocalized electrons' spins.

Diamagnetic materials are those in which the electron motions are such that they produce net zero magnetic moment in the absence of any magnetic field. Typically these are atoms with closed or filled outer electron shells.

Examples of such materials are inert gases, hydrogen, many metals (e.g. Ag, Au, Cu etc.), most non-metals (e.g. Si) and many organic compounds such as polymers.

Imagine a circular orbit of radius, r, around an atom with its center coinciding that of the atom. Now, we turn on the magnetic field in its vicinity.

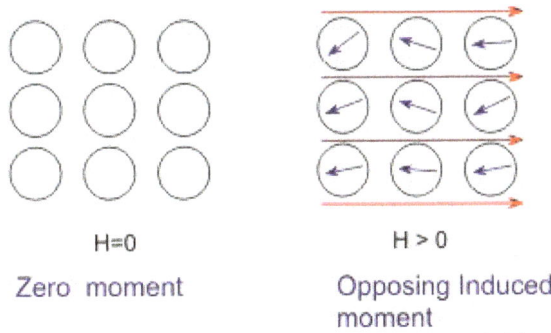

H=0   H > 0

Zero moment   Opposing Induced
                                                  moment

Representation of a diamagnetic material in the absence of a magnetic field and when a field is applied,
note than when field is applied, induced moments oppose the field.

Hence, according to Faraday's law, as the magnetic field changes, it generates an electric field by magnetic induction. The electric field, E, tangent to the circular path is given as

$$E.2\pi r = -\frac{d(B.\pi r^2)}{dt}$$ OR

$$E = -\frac{r}{2}\frac{dB}{dt}$$

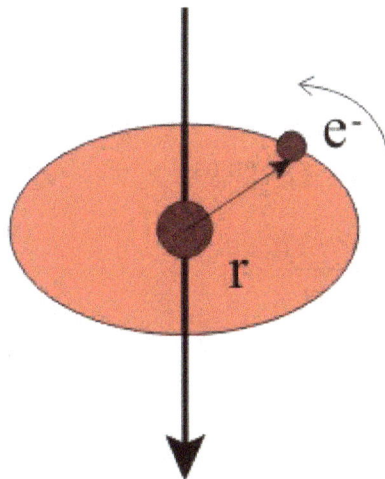

Motion of an electron in an atom's orbit

This electric field produces a torque equivalent to -eE.r (i.e. F.r) which must be equal to the rate of change of angular momentum, J, i.e

$$\frac{dJ}{dt} = e.\frac{r}{2}\frac{dB}{dt}.r$$

Two minuses cancel each other. Now integrating with respect to time with zero field, we get

$$\Delta J = \left(\frac{er^2}{2}\right)B$$

This expression represents the extra angular momentum provided to the electrons when the field is applied.

Now, since the motion of electron is taking place in the orbits, the change in magnetic moment ($\Delta\mu_m$) which is orbital in nature is given as

Replacing $B = \mu_0 H$, we get

$$\Delta\mu_m = -\left(\frac{e^2 r^2 \mu_0}{4m}\right) H$$

For atoms with spherical symmetry, $\langle r^2 \rangle = (2/3).\langle \bar{r}^2 \rangle$, hence,

$$\Delta\mu_m = -\frac{e^2 \bar{r}^2 \mu_0}{6m} H$$

So, here we have a magnetic moment which is negative in sign to the magnetic field strength because it opposes the applied field. This magnetic moment is the moment for one electron.

So, if there are N electrons per unit volume, then magnetization, M, is given as

$$M = -\frac{Ne^2 \bar{r}^2 \mu_2}{6m} H$$

Hence magnetic susceptibility χ_m is given as

$$X_m = \frac{M}{H} = -\frac{Ne^2 \bar{r}^2 \mu_0}{6m}$$

This equation shows the opposite nature of the magnetic susceptibility of the diamagnetic behaviour. Remember that diamagnetism is something which is present in all materials except that many materials also have other effects which completely overshadow the diamagnetic effect.

Paramagnetism

Paramagnetism is a form of magnetism whereby certain materials are weakly attracted by an externally applied magnetic field, and form internal, induced magnetic fields in the direction of the applied magnetic field. In contrast with this behavior, diamagnetic materials are repelled by magnetic fields and form induced magnetic fields in the direction opposite to that of the applied magnetic field. Paramagnetic materials include most chemical elements and some compounds; they have a relative magnetic permeability greater than or equal to 1 (i.e., a small positive magnetic susceptibility) and hence are attracted to magnetic fields. The magnetic moment induced by the applied field is linear in the field strength and rather weak. It typically requires a sensitive analytical balance to detect the effect and modern measurements on paramagnetic materials are often conducted with a SQUID magnetometer.

Paramagnetism is due to the presence of unpaired electrons in the material, so all atoms with incompletely filled atomic orbitals are paramagnetic. Due to their spin, unpaired electrons have a magnetic dipole moment and act like tiny magnets. An external magnetic field causes the electrons' spins to align parallel to the field, causing a net attraction. Paramagnetic materials include aluminum, oxygen, titanium, and iron oxide (FeO).

Unlike ferromagnets, paramagnets do not retain any magnetization in the absence of an externally applied magnetic field because thermal motion randomizes the spin orientations. (Some paramagnetic materials retain spin disorder even at absolute zero, meaning they are paramagnetic in the ground state, i.e. in the absence of thermal motion.) Thus the total magnetization drops to zero when the applied field is removed. Even in the presence of the field there is only a small induced magnetization because only a small fraction of the spins will be oriented by the field. This fraction is proportional to the field strength and this explains the linear dependency. The attraction experienced by ferromagnetic materials is non-linear and much stronger, so that it is easily observed, for instance, in the attraction between a refrigerator magnet and the iron of the refrigerator itself.

Relation to Electron Spins

Constituent atoms or molecules of paramagnetic materials have permanent magnetic moments (dipoles), even in the absence of an applied field. The permanent moment generally is due to the spin of unpaired electrons in atomic or molecular electron orbitals. In pure paramagnetism, the dipoles do not interact with one another and are randomly oriented in the absence of an external field due to thermal agitation, resulting in zero net magnetic moment. When a magnetic field is applied, the dipoles will tend to align with the applied field, resulting in a net magnetic moment in the direction of the applied field. In the classical description, this alignment can be understood to occur due to a torque being provided on the magnetic moments by an applied field, which tries to align the dipoles parallel to the applied field. However, the true origins of the alignment can only be understood via the quantum-mechanical properties of spin and angular momentum.

If there is sufficient energy exchange between neighbouring dipoles they will interact, and may spontaneously align or anti-align and form magnetic domains, resulting in ferromagnetism (permanent magnets) or antiferromagnetism, respectively. Paramagnetic behavior can also be observed in ferromagnetic materials that are above their Curie temperature, and in antiferromagnets above their Néel temperature. At these temperatures, the available thermal energy simply overcomes the interaction energy between the spins.

In general, paramagnetic effects are quite small: the magnetic susceptibility is of the order of 10^{-3} to 10^{-5} for most paramagnets, but may be as high as 10^{-1} for synthetic paramagnets such as ferrofluids.

Delocalization

Selected Pauli-paramagnetic metals	
Material	**Magnetic susceptibility, $\chi_v \left[10^{-5} \right]$**
Tungsten	6.8
Cesium	5.1
Aluminium	2.2
Lithium	1.4
Magnesium	1.2
Sodium	0.72

In conductive materials the electrons are delocalized, that is, they travel through the solid more or less as free electrons. Conductivity can be understood in a band structure picture as arising from the incomplete filling of energy bands. In an ordinary nonmagnetic conductor the conduction band is identical for both spin-up and spin-down electrons. When a magnetic field is applied, the conduction band splits apart into a spin-up and a spin-down band due to the difference in magnetic potential energy for spin-up and spin-down electrons. Since the Fermi level must be identical for both bands, this means that there will be a small surplus of the type of spin in the band that moved downwards. This effect is a weak form of paramagnetism known as *Pauli paramagnetism*.

The effect always competes with a diamagnetic response of opposite sign due to all the core electrons of the atoms. Stronger forms of magnetism usually require localized rather than itinerant electrons. However, in some cases a band structure can result in which there are two delocalized sub-bands with states of opposite spins that have different energies. If one subband is preferentially filled over the other, one can have itinerant ferromagnetic order. This situation usually only occurs in relatively narrow (d-)bands, which are poorly delocalized.

s and p Electrons

Generally, strong delocalization in a solid due to large overlap with neighboring wave functions means that there will be a large Fermi velocity; this means that the number of electrons in a band is less sensitive to shifts in that band's energy, implying a weak magnetism. This is why s- and p-type metals are typically either Pauli-paramagnetic or as in the case of gold even diamagnetic. In the latter case the diamagnetic contribution from the closed shell inner electrons simply wins from the weak paramagnetic term of the almost free electrons.

d and f Electrons

Stronger magnetic effects are typically only observed when d or f electrons are involved. Particularly the latter are usually strongly localized. Moreover, the size of the magnetic moment on a lanthanide atom can be quite large as it can carry up to 7 unpaired electrons in the case of gadolinium(III) (hence its use in MRI). The high magnetic moments associated with lanthanides is one reason why superstrong magnets are typically based on elements like neodymium or samarium.

Molecular Localization

Of course the above picture is a *generalization* as it pertains to materials with an extended lattice rather than a molecular structure. Molecular structure can also lead to localization of electrons. Although there are usually energetic reasons why a molecular structure results such that it does not exhibit partly filled orbitals (i.e. unpaired spins), some non-closed shell moieties do occur in nature. Molecular oxygen is a good example. Even in the frozen solid it contains di-radical molecules resulting in paramagnetic behavior. The unpaired spins reside in orbitals derived from oxygen p wave functions, but the overlap is limited to the one neighbor in the O_2 molecules. The distances to other oxygen atoms in the lattice remain too large to lead to delocalization and the magnetic moments remain unpaired.

Curie's Law

For low levels of magnetization, the magnetization of paramagnets follows what is known as Curie's law, at least approximately. This law indicates that the susceptibility, χ, of paramagnetic materials is inversely proportional to their temperature, i.e. that materials become more magnetic at lower temperatures. The mathematical expression is:

$$\mathbf{M} = \chi \mathbf{H} = \frac{C}{T} \mathbf{H}$$

where:

    $\mathbf{M}$ is the resulting magnetization

    χ is the magnetic susceptibility

    $\mathbf{H}$ is the auxiliary magnetic field, measured in amperes/meter

    T is absolute temperature, measured in kelvins

    C is a material-specific Curie constant

A derivation of this law:

Curie's law is valid under the commonly encountered conditions of low magnetization ($\mu_B H \lesssim k_B T$), but does not apply in the high-field/low-temperature regime where saturation of magnetization occurs ($\mu_B H \gtrsim k_B T$) and magnetic dipoles are all aligned with the applied field. When the dipoles are aligned, increasing the external field will not increase the total magnetization since there can be no further alignment.

For a paramagnetic ion with noninteracting magnetic moments with angular momentum J, the Curie constant is related the individual ions' magnetic moments,

$$C = \frac{N_A}{3k_B} \mu_{eff}^2 \ \text{ where } \ \mu_{eff} = g_J \mu_B \sqrt{J(J+1)}$$

The parameter μ_{eff} is interpreted as the effective magnetic moment per paramagnetic ion. If one uses a classical treatment with molecular magnetic moments represented as discrete magnetic dipoles, μ, a Curie Law expression of the same form will emerge with μ appearing in place of μ_{eff}.

Curie's Law can be derived by considering a substance with noninteracting magnetic moments with angular momentum J. If orbital contributions to the magnetic moment are negligible (a common case), then in what follows J = S. If we apply a magnetic field along what we choose to call the z-axis, the energy levels of each paramagnetic center will experience Zeeman splitting of its energy levels, each with a z-component labeled by M_J (or just M_S for the spin-only magnetic case). Applying semiclassical Boltzmann statistics, the molar magnetization of such a substance is

$$N_A \overline{m} = \frac{N_A \sum\limits_{M_J=-J}^{J} \mu_{M_J} e^{-E_{M_J}/k_B T}}{\sum\limits_{M_J=-J}^{J} e^{-E_{M_J}/k_B T}} = \frac{N_A \sum\limits_{M_J=-J}^{J} M_J g_J \mu_B e^{M_J g_J \mu_B H/k_B T}}{\sum\limits_{M_J=-J}^{J} e^{M_J g_J \mu_B H/k_B T}}.$$

Where μ_{M_J} is the z-component of the magnetic moment for each Zeeman level, so $\mu_{M_J} = M_J g_J \mu_B - \mu_B$ is called the Bohr magneton and g_J is the Landé g-factor, which reduces to the free-electron g-factor, g_s when J = S. (in this treatment, we assume that the x- and y-components of the magnetization, averaged over all molecules, cancel out because the field applied along the z-axis leave them randomly oriented.) The energy of each Zeeman level is $E_{M_J} = -M_J g_J \mu_B H$. For temperatures over a few K, $M_J g_J \mu_B H / k_B T \ll 1$, and we can apply the approximation

$$e^{M_J g_J \mu_B H/k_B T} \simeq 1 + M_J g_J \mu_B H / k_B T$$

$$\overline{m} = \frac{\sum\limits_{M_J=-J}^{J} M_J g_J \mu_B e^{M_J g_J \mu_B H/k_B T}}{\sum\limits_{M_J=-J}^{J} e^{M_J g_J \mu_B H/k_B T}} \simeq g_J \mu_B \frac{\sum\limits_{M_J=-J}^{J} M_J \left(1 + M_J g_J \mu_B H / k_B T\right)}{\sum\limits_{M_J=-J}^{J} \left(1 + M_J g_J \mu_B H / k_B T\right)} = \frac{g_J^2 \mu_B^2 H}{k_B T} \frac{\sum\limits_{-J}^{J} M_J^2}{\sum\limits_{M_J=-J}^{J} (1)},$$

which yields:

$$\bar{m} = \frac{g_J^2 \mu_B^2 H}{3k_B T} J(J+1).$$ The molar bulk magnetization is then

$$M = N_A \bar{m} = \frac{N_A}{3k_B T}\left[g_J^2 J(J+1)\mu_B^2 \right] H \quad,$$

and the molar susceptibility is given by

$$\chi_m = \frac{\partial M}{\partial H} = \frac{N_A}{3k_B T}\mu_{\mathit{eff}}^2 \quad; \quad and \quad \mu_{\mathit{eff}} = g_J \sqrt{J(J+1)}\mu_B.$$

When orbital angular momentum contributions to the magnetic moment are small, as occurs for most organic radicals or for octahedral transition metal complexes with d³ or high-spin d⁵ configurations, the effective magnetic moment takes the form (g_e = 2.0023... ≈ 2),

$$\mu_{\mathit{eff}} \simeq 2\sqrt{S(S+1)}\mu_B = \sqrt{n(n+2)}\mu_B,$$

where n is the number of unpaired electrons. In other transition metal complexes this yields a useful, if somewhat cruder, estimate.

Examples of Paramagnets

Materials that are called "paramagnets" are most often those that exhibit, at least over an appreciable temperature range, magnetic susceptibilities that adhere to the Curie or Curie–Weiss laws. In principle any system that contains atoms, ions, or molecules with unpaired spins can be called a paramagnet, but the interactions between them need to be carefully considered.

Systems with Minimal Interactions

The narrowest definition would be: a system with unpaired spins that *do not interact* with each other. In this narrowest sense, the only pure paramagnet is a dilute gas of monatomic hydrogen atoms. Each atom has one non-interacting unpaired electron. Of course, the latter could be said about a gas of lithium atoms but these already possess two paired core electrons that produce a diamagnetic response of opposite sign. Strictly speaking Li is a mixed system therefore, although admittedly the diamagnetic component is weak and often neglected. In the case of heavier elements the diamagnetic contribution becomes more important and in the case of metallic gold it dominates the properties. Of course, the element hydrogen is virtually never called 'paramagnetic' because the monatomic gas is stable only at extremely high temperature; H atoms combine to form molecular H_2 and in so doing, the magnetic moments are lost (*quenched*), because of the spins pair. Hydrogen is therefore *diamagnetic* and the same holds true for many other elements. Although the electronic configuration of the individual atoms (and ions) of most elements contain unpaired spins, they are not necessarily paramagnetic, because at ambient temperature quenching is very much the rule rather than the exception. The quenching tendency is weakest for f-electrons because f (especially $4f$) orbitals are radially contracted and they overlap only weakly with orbitals on adjacent atoms. Consequently, the lanthanide elements with incompletely filled 4f-orbitals are paramagnetic or magnetically ordered.

μ_{eff} values for typical d³ and d⁵ transition metal complexes.	
Material	μ_{eff}/μ_B
$[Cr(NH_3)_6]Br_3$	3.77
$K_3[Cr(CN)_6]$	3.87
$K_3[MoCl_6]$	3.79
$K_4[V(CN)_6]$	3.78
$[Mn(NH_3)_6]Cl_2$	5.92
$(NH_4)_2[Mn(SO_4)_2]\cdot 6H_2O$	5.92
$NH_4[Fe(SO_4)_2]\cdot 12H_2O$	5.89

Thus, condensed phase paramagnets are only possible if the interactions of the spins that lead either to quenching or to ordering are kept at bay by structural isolation of the magnetic centers. There are two classes of materials for which this holds:

- Molecular materials with a (isolated) paramagnetic center.

  o Good examples are coordination complexes of d- or f-metals or proteins with such centers, e.g. myoglobin. In such materials the organic part of the molecule acts as an envelope shielding the spins from their neighbors.

  o Small molecules can be stable in radical form, oxygen O_2 is a good example. Such systems are quite rare because they tend to be rather reactive.

- Dilute systems.

  o Dissolving a paramagnetic species in a diamagnetic lattice at small concentrations, e.g. Nd^{3+} in $CaCl_2$ will separate the neodymium ions at large enough distances that they do not interact. Such systems are of prime importance for what can be considered the most sensitive method to study paramagnetic systems: EPR.

Systems with Interactions

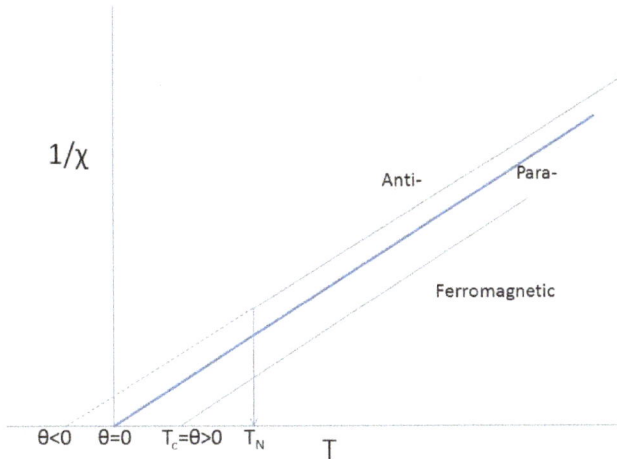

Idealized Curie–Weiss behavior; N.B. $T_C=\theta$, but T_N is not θ. Paramagnetic regimes are denoted by solid lines. Close to T_N or T_C the behavior usually deviates from ideal.

As stated above, many materials that contain d- or f-elements do retain unquenched spins. Salts of such elements often show paramagnetic behavior but at low enough temperatures the magnetic moments may order. It is not uncommon to call such materials 'paramagnets', when referring to their paramagnetic behavior above their Curie or Néel-points, particularly if such temperatures are very low or have never been properly measured. Even for iron it is not uncommon to say that *iron becomes a paramagnet* above its relatively high Curie-point. In that case the Curie-point is seen as a phase transition between a ferromagnet and a 'paramagnet'. The word paramagnet now merely refers to the linear response of the system to an applied field, the temperature dependence of which requires an amended version of Curie's law, known as the Curie–Weiss law:

$$\mathbf{M} = \frac{C}{T-\theta}\mathbf{H}$$

This amended law includes a term θ that describes the exchange interaction that is present albeit overcome by thermal motion. The sign of θ depends on whether ferro- or antiferromagnetic interactions dominate and it is seldom exactly zero, except in the dilute, isolated cases mentioned above.

Obviously, the paramagnetic Curie–Weiss description above T_N or T_C is a rather different interpretation of the word "paramagnet" as it does *not* imply the *absence* of interactions, but rather that the magnetic structure is random in the absence of an external field at these sufficiently high temperatures. Even if θ is close to zero this does not mean that there are no interactions, just that the aligning ferro- and the anti-aligning antiferromagnetic ones cancel. An additional complication is that the interactions are often different in different directions of the crystalline lattice (anisotropy), leading to complicated magnetic structures once ordered.

Randomness of the structure also applies to the many metals that show a net paramagnetic response over a broad temperature range. They do not follow a Curie type law as function of temperature however, often they are more or less temperature independent. This type of behavior is of an itinerant nature and better called Pauli-paramagnetism, but it is not unusual to see, for example, the metal aluminium called a "paramagnet", even though interactions are strong enough to give this element very good electrical conductivity.

Superparamagnets

Some materials show induced magnetic behavior that follows a Curie type law but with exceptionally large values for the Curie constants. These materials are known as superparamagnets. They are characterized by a strong ferromagnetic or ferrimagnetic type of coupling into domains of a limited size that behave independently from one another. The bulk properties of such a system resembles that of a paramagnet, but on a microscopic level they are ordered. The materials do show an ordering temperature above which the behavior reverts to ordinary paramagnetism (with interaction). Ferrofluids are a good example, but the phenomenon can also occur inside solids, e.g., when dilute paramagnetic centers are introduced in a strong itinerant medium of ferromagnetic coupling such as when Fe is substituted in $TlCu_2Se_2$ or the alloy AuFe. Such systems contain ferromagnetically coupled clusters that freeze out at lower temperatures. They are also called mictomagnets.

In paramagnetic materials, atoms have a permanent non-zero net magnetic moment due to the sum of orbital and spin magnetic moments. However at room temperature, in paramagnetic materials, thermal energy causes random distribution of magnetic moments and hence net magnetization appears to be zero for the whole material (not just an atom).

Upon application of a field, the moments tend to align up in the direction of the field overcoming the thermal barrier and giving a net positive magnetic moment in the direction of the applied field. The susceptibilities of these materials are usually very small, 10^{-3} to 10^{-6}.

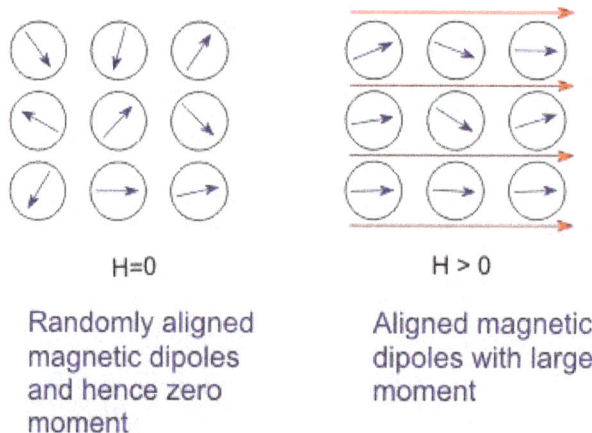

H=0

H > 0

Randomly aligned magnetic dipoles and hence zero moment

Aligned magnetic dipoles with large moment

Schematic representation of spins in a paramagnetic solid

In most solids, only spin paramagnetism is observed, since the electron orbits are considered as coupled to the lattice (the orbital moments are quenched) and hence do not contribute significantly to the magnetic moment. Qualitatively, it is due to the electric field generated by the surrounding ions in the solid. Because of these electric fields, the orbitals are strongly coupled to the lattice and hence cannot reorient themselves along the field and therefore, do not contribute towards the magnetic moment. On the other hand, electron spins are weakly coupled and hence form a major part of the magnetic moment.

Spin paramagnetism is observed in materials with unfilled d-shells which follow Hund's rule which says that in the partially filled d-orbitals, the moment is maximized by the first filling all the spin states of one direction followed by filling in other direction.

For example, Ni, with an atomic number of 28 has an electronic configuration of $1s^2$, $2s^2$, $2p^6$, $3s^2$, $3p^6$, $4s^2$ and $3d^8$. The filling in of partially occupied d-orbital is achieved as follows:

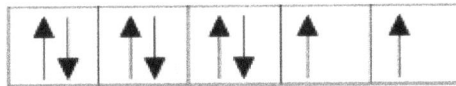

As a result, each Ni atom has a net magnetization of $2\mu_B$.

Exception to this may be rare-earth elements and their derivatives which have deep-lying 4f- electrons. These are shielded by the outer electrons from the crystal field and as result they show both spin and orbital paramagnetism.

Susceptibility (from orbital moment) in paramagnetic material shows temperature dependence as governed by Curie's law and is given as

$$\chi_{para}^{orbit} = \frac{M}{H} = \frac{N\mu_m^2\mu_o}{3kT} = \frac{C}{T}$$

where N is the number of atoms per unit volume, μ_m is the magnetic moment of an atom and C is Curie constant. Curie's law states that susceptibility of a paramegntic material is inversely proportional to temperature.

Here we assume that each atom has a magnetic moment μ_m whose magnitude is the same but the direction can be random. The magnetic energy in a field B would be $\mu_m.B = -\mu_m B\cos\alpha$ where α is the angle between the moment and field.

Boltzmann statistics in classical thermodynamics gives the probability of having any angle i.e. of occupying any energy as $\exp(-\mu_m H.\cos\alpha/kT)$. As a result, as you can also notice that it is more plausible to have moment closer to 0° than 180° with respect to the applied field, simply because that leads to smaller energy than those at 180°.

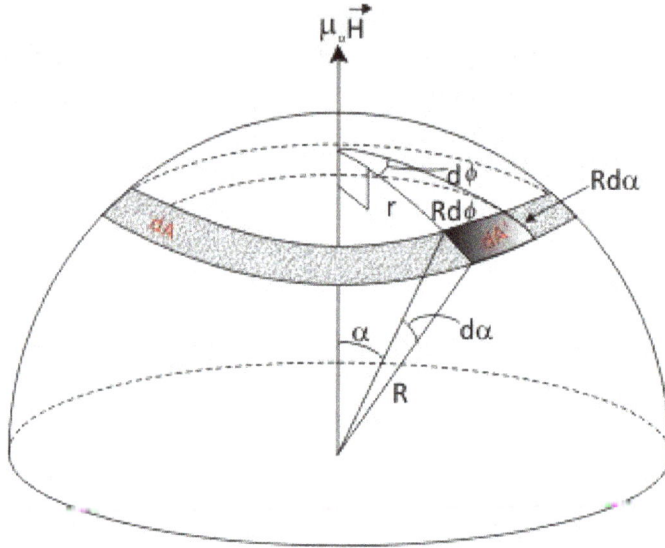

Representation of magnetic moment in a paramagnetic solid with respect to the applied field

Consider a smaller number of dipoles dn making an angle a with respect to the applied field and penetrating an area dA as shown in figure which is proportional to exp(-μ_mH-cosα/kT)

Considering a unit radius of the sphere (R=1), the area dA can be worked out as

$$dA = 2\pi r^2 \sin \alpha \, d\alpha$$

Hence dn can be expressed as

$$dn = C.dA.\exp\left(-\mu_m{}^{H \cos \alpha} / {}_{kT}\right)$$

or

$$dn = C.2\pi r^2 \sin \alpha \, d\alpha \exp\left(-\mu_m{}^{H \cos \alpha} / {}_{kT}\right)$$

where C is a constant.

Assuming that $\gamma = \mu_m H/kT$, we can integrate the above equation for a between 0 to 180° which will yield the total number of dipoles i.e.

$$N = \int_0^\pi C.2\pi r^2 \sin \alpha \, d\alpha \exp(-\gamma \cos \alpha) \, d\alpha$$

Now, the total magnetic moment along the direction of magnetic field can be expressed multiplication of number dipoles multiplied by the magnetic moment of each dipole along the field direction ($\mu_m$cosα) and is given as

$$M_{tot} = \int_0^\pi \mu_m \cos \alpha.C.2\pi r^2 \sin \alpha \, d\alpha \exp\left(-\mu_m{}^{H \cos \alpha} / {}_{kT}\right)$$

Dividing equations yields the average magnetization, M_{avj}, as

$$M_{avj} = \frac{\int_0^\pi \mu_m \cos\alpha.C.2\pi r^2 \sin\alpha\, d\alpha \exp\left(-\mu_m^{H\cos\alpha}/kT\right)}{\int_0^\pi C.2\pi r^2 \sin\alpha\, d\alpha \exp(-\gamma\cos\alpha)d\alpha}$$

$$= \frac{\mu_m \int_0^\pi \cos\alpha.C.2\pi r^2 \sin\alpha\, d\alpha \exp(-\gamma\cos\alpha)d\alpha}{\int_0^\pi \sin\alpha\, d\alpha \exp(-\gamma\cos\alpha)d\alpha}$$

Now, following orientation polarization, we can show that for reasonably small magnetic field, moments are parallel to applied field and magnetization M can be expressed as

$$M = N\mu_m \cdot \left\{ \coth\left(\frac{\mu_m B}{kT}\right) - \frac{kT}{\mu_m B} \right\} = N\mu_m L(\beta)$$

where $\beta = (\mu_m B)(kT)$ and $L(\beta) = \cot h\beta - 1/\beta$ is called Langevin function. For most cases i.e. when $\mu_m B << kT$, we can assume $L(\beta) = \beta/3$. Using this approximation, we can write equation (28) as

$$M = \frac{N\mu_m^2 B}{3kT} = \frac{N\mu_m^2 \mu_\circ H}{3kT}$$

OR

$$\chi_{para} = \frac{M}{H} = \frac{C}{T}$$

where Curie constant $C = (N\mu_m^2 \mu_\circ)/3k$.

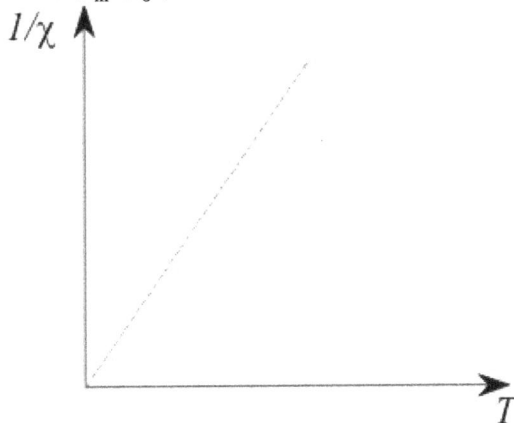

Susceptibility vs temperature plot for a paramagnetic material

Equations show that induced magnetization is proportional to the applied field and is larger at lower temperature. The same result can also be shown using quantum mechanism which is not a part of this course.

Quantum mechanics, for any characteristic number j (could be combined for both orbital and spin moments or only s for spin moments), also gives

$$\chi_{para} = g^2 j(j+1)\frac{N\mu_B^2\mu_\circ}{3kT} = \frac{C}{T}$$

which is similar to $\mu_m = \mu_{eff} = g.\sqrt{j(j+1)}.\mu_B$. Here, g is Landé-g factor.

The theoretically derived and experimentally obtained values of magnetic moment for a few selected magnetic ions are given below and here, you can see while for rare earth ions total magnetic moment value gives a much better estimate, for transition elements, spin only magnetism in a better estimate

Ion	Configuration	Calculated		Measured
		$g.\sqrt{j(j+1)}$	$g.\sqrt{s(s+1)}$	$d(\mu_m/\mu_B)$
Rare Earths				
Ce³⁺	$4f^1 5s^2 5p^6$	2.54	-	2.4
Pr³⁺	$4f^2 5s^2 5p^6$	3.58	-	3.5
Nd³⁺	$4f^3 5s^2 5p^6$	3.62	-	3.5
Transition Elements				
Mn³⁺	$3d^4$	0.00	4.90	4.9
Mn²⁺, Fe³⁺	$3d^5$	5.92	5.92	5.9
Fe²⁺	$3d^6$	6.70	4.90	5.4
Co²⁺	$3d^7$	6.63	3.87	4.8
Ni²⁺	$3d^8$	5.59	2.83	3.2

Ferrite (Magnet)

A ferrite is a type of ceramic compound composed of iron(III) oxide (Fe_2O_3) combined chemically with one or more additional metallicelements. They are both electrically nonconductive and ferrimagnetic, meaning they can be magnetized or attracted to a magnet. Ferrites can be divided into two families based on their magnetic coercivity, their resistance to being demagnetized. *Hard ferrites* have high coercivity, hence they are difficult to demagnetize. They are used to make magnets, for devices such as refrigerator magnets, loudspeakers and small electric motors. *Soft ferrites* have low coercivity. They are used in the electronics industry to make ferrite cores for inductors and transformers, and in various microwave components. Ferrite compounds have extremely low cost, being made of iron oxide (i.e. rusted iron), and also have excellent corrosion resistance. They are very stable and difficult to demagnetize, and can be made with both high and low coercive forces. Yogoro Kato and Takeshi Takei of the Tokyo Institute of Technology synthesized the first ferrite compounds in 1930.

Composition and Properties

Ferrites are usually non-conductive ferrimagnetic ceramic compounds derived from iron oxides such as hematite (Fe_2O_3) or magnetite (Fe_3O_4) as well as oxides of other metals. Ferrites are, like most of the other ceramics, hard and brittle.

Many ferrites are spinels with the formula AB_2O_4, where A and B represent various metal cations, usually including iron Fe. Spinel ferrites usually adopt a crystal motif consisting of cubic close-packed (fcc) oxides (O^{2-}) with A cations occupying one eighth of the tetrahedral holes and B cations occupying half of the octahedral holes. If one eighth of the tetrahedral holes are taken by B cation, then one fourth of the octahedral sites are occupied by A cation and the other one fourth by B cation and it is called the inverse spinel structure. It is also possible to have mixed structure spinel ferrites with formula $[M^{2+}_{1-\delta}Fe^{3+}_{\delta}][M^{2+}_{\delta}Fe^{3+}_{2-\delta}]O_4$ where δ is the degree of inversion.

The magnetic material known as "ZnFe" has the formula $ZnFe_2O_4$, with Fe^{3+} occupying the octahedral sites and Zn^{2+} occupy the tetrahedral sites, it is an example of normal structure spinel ferrite.

Some ferrites have hexagonal crystal structure, like barium and strontium ferrites $BaFe_{12}O_{19}$ ($BaO:6Fe_2O_3$) and $SrFe_{12}O_{19}$ ($SrO:6Fe_2O_3$).

In terms of their magnetic properties, the different ferrites are often classified as "soft" or "hard", which refers to their low or high magnetic coercivity, as follows.

Soft Ferrites

Various ferrite cores used to make small transformers and inductors

Ferrites that are used in transformer or electromagnetic cores contain nickel, zinc, and/or manganese compounds. They have a low coercivity and are called soft ferrites. The low coercivity means the material's magnetization can easily reverse direction without dissipating much energy (hysteresis losses), while the material's high resistivity prevents eddy currents in the core, another source of energy loss. Because of their comparatively low losses at high frequencies, they are extensively used in the cores of RF transformers and inductors in applications such as switched-mode power supplies and loopstick antennas used in AM radios.

The most common soft ferrites are:

- Manganese-zinc ferrite (MnZn, with the formula $Mn_aZn_{(1-a)}Fe_2O_4$). MnZn have higher permeability and saturation induction than NiZn.

- Nickel-zinc ferrite (NiZn, with the formula $Ni_aZn_{(1-a)}Fe_2O_4$). NiZn ferrites exhibit higher resistivity than MnZn, and are therefore more suitable for frequencies above 1 MHz.

For applications below 5 MHz, MnZn ferrites are used; above that, NiZn is the usual choice. The exception is with common mode inductors, where the threshold of choice is at 70 MHz.

Hard Ferrites

In contrast, permanent ferrite magnets are made of hard ferrites, which have a high coercivity and high remanence after magnetization. Iron oxide and barium or strontium carbonate are used in manufacturing of hard ferrite magnets. The high coercivity means the materials are very resistant to becoming demagnetized, an essential characteristic for a permanent magnet. They also have high magnetic permeability. These so-called *ceramic magnets* are cheap, and are widely used in household products such as refrigerator magnets. The maximum magnetic field B is about 0.35 tesla and the magnetic field strength H is about 30 to 160 kiloampere turns per meter (400 to 2000 oersteds). The density of ferrite magnets is about 5 g/cm³.

The most common hard ferrites are:

- Strontium ferrite, $SrFe_{12}O_{19}$ ($SrO \cdot 6Fe_2O_3$), used in small electric motors, micro-wave devices, recording media, magneto-optic media, telecommunication and electronic industry.

- Barium ferrite, $BaFe_{12}O_{19}$ ($BaO \cdot 6Fe_2O_3$), a common material for permanent magnet applications. Barium ferrites are robust ceramics that are generally stable to moisture and corrosion-resistant. They are used in e.g. loudspeaker magnets and as a medium for magnetic recording, e.g. on magnetic stripe cards.

- Cobalt ferrite, $CoFe_2O_4$ ($CoO \cdot Fe_2O_3$), used in some media for magnetic recording.

Production

Ferrites are produced by heating a mixture of finely-powdered precursors pressed into a mold. During the heating process, calcination of carbonates occurs:

$$MCO_3 \rightarrow MO + CO_2$$

The oxides of barium and strontium are typically supplied as their carbonates, $BaCO_3$ or $SrCO_3$. The resulting mixture of oxides undergoes sintering. Sintering is a high temperature process similar to the firing of ceramic ware.

Afterwards, the cooled product is milled to particles smaller than 2 μm, small enough that each particle consists of a single magnetic domain. Next the powder is pressed into a shape, dried, and re-sintered. The shaping may be performed in an external magnetic field, in order to achieve a preferred orientation of the particles (anisotropy).

Small and geometrically easy shapes may be produced with dry pressing. However, in such a process small particles may agglomerate and lead to poorer magnetic properties compared to the wet pressing process. Direct calcination and sintering without re-milling is possible as well but leads to poor magnetic properties.

Electromagnets are pre-sintered as well (pre-reaction), milled and pressed. However, the sintering takes place in a specific atmosphere, for instance one with an oxygen shortage. The chemical composition and especially the structure vary strongly between the precursor and the sintered product.

To allow efficient stacking of product in the furnace during sintering and prevent parts sticking together, many manufacturers separate ware using ceramic powder separator sheets. These sheets are available in various materials such as alumina, zirconia and magnesia. They are also available in fine, medium and coarse particle sizes. By matching the material and particle size to the ware being sintered, surface damage and contamination can be reduced while maximizing furnace loading.

Uses

Ferrite cores are used in electronic inductors, transformers, and electromagnets where the high electrical resistance of the ferrite leads to very low eddy current losses. They are commonly seen as a lump in a computer cable, called a ferrite bead, which helps to prevent high frequency electrical noise (radio frequency interference) from exiting or entering the equipment.

Early computer memories stored data in the residual magnetic fields of hard ferrite cores, which were assembled into arrays of *core memory*. Ferrite powders are used in the coatings of magnetic recording tapes. One such type of material is iron (III) oxide.

Ferrite particles are also used as a component of radar-absorbing materials or coatings used in stealth aircraft and in the absorption tiles lining the rooms used for electromagnetic compatibility measurements.

Most common radio magnets, including those used in loudspeakers, are ferrite magnets. Ferrite magnets have largely displaced Alnico magnets in these applications.

It is a common magnetic material for electromagnetic instrument pickups.

Ferrite nanoparticles exhibit superparamagnetic properties.

History

Yogoro Kato and Takeshi Takei of the Tokyo Institute of Technology synthesized the first ferrite compounds in 1930. This led to the founding of TDK Corporation in 1935, to manufacture the material.

Barium hexaferrite ($BaFe_{12}O_{19}$) was discovered in 1950 at the Philips Natuurkundig Laboratorium (*Philips Physics Laboratory*). The discovery was somewhat accidental—due to a mistake by an assistant who was supposed to be preparing a sample of hexagonal lanthanum ferrite for a team investigating its use as a semiconductor material. On discovering that it was actually a magnetic material, and confirming its structure by X-ray crystallography, they passed it on to the magnetic research group. Barium hexaferrite has both high coercivity (170 kA/m) and low raw material costs. It was developed as a product by Philips Industries (Netherlands) and from 1952 was marketed under the trade name *Ferroxdure*. The low price and good performance led to a rapid increase in the use permanent magnets.

In the 1960s Philips developed strontium hexaferrite ($SrFe_{12}O_{19}$), with better properties than barium hexaferrite. Barium and strontium hexaferrite dominate the market due to their low costs. However other materials have been found with improved properties. $BaFe^{2+}{}_2Fe^{3+}{}_{16}O_{27}$ came in 1980 and $Ba_2ZnFe_{18}O_{23}$ came in 1991.

Ferromagnetism

Basic Characteristics

In addition to permanent magnetic moments as contained in paramagnetic materials, ferromagnetic materials consist of ordered regions or domains of single orientation of magnetic moment giving rise to large finite magnetization in the absence of a magnetic field, much like polarization in ferroelectric materials. This phenomenon is observed below a critical temperature called as Curie Temperature, above which the material behave like a paramagnetic material.

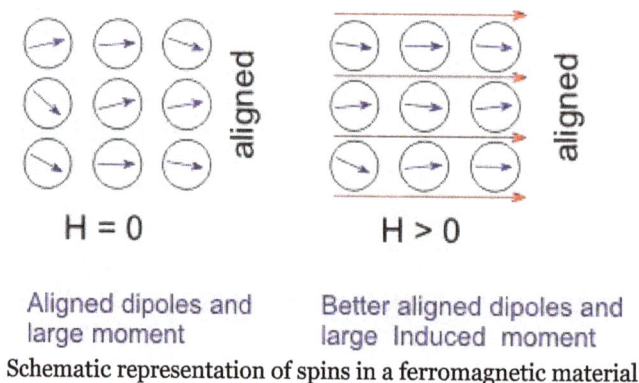

Aligned dipoles and large moment

Better aligned dipoles and large Induced moment

Schematic representation of spins in a ferromagnetic material

When a varying magnetic field is applied to a ferromagnetic material, the material exhibits a ferroelectric-like hysteresis loop between magnetization and the magnetic field as shown below. In this figure, notice how the domain structure changes with field in the initial stages of magnetization.

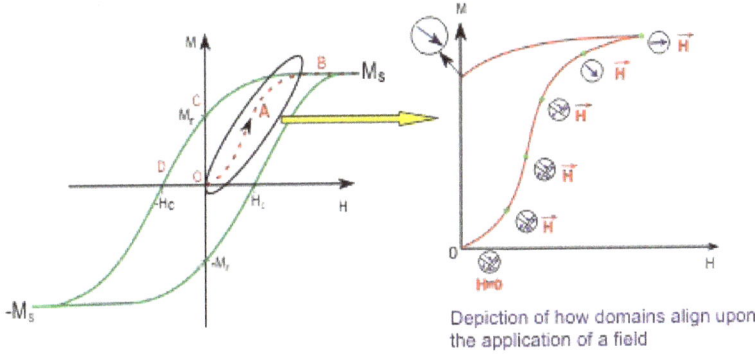

Depiction of how domains align upon the application of a field

Ferromagnetic hysteresis loops

Most of the ferromagnetic materials are elemental metals such as iron, nickel, cobalt etc. However some oxides such as chrmoimum oxide, CrO_2, are ferromagnetic oxides. These oxides also tend to be conducting and behave like metals.

Antiferromagnetic Materials

These are materials in which electron spins associated with magnetic atoms at particular crystallographic sites are ordered yet oriented with respect to each other in such a manner that their net magnetization is equal to zero. This is the case below a particular temperature, called as Néel temperature (T_N) above which the material behaves as a paramagnet.

Examples include metallic manganese, chromium, various transition metal oxides such as manganese oxide (MnO), forms of iron oxide (Fe_2O_3), multiferroic perovskites like bismuth ferrite ($BiFeO_3$). For example in MnO, since O is not a magnetic ion, the antiparallel spin arrangement of Mn^{2+}, the magnetic ion, in two sites gives rise to zero magnetization. Below is the crystal structure of MnO, drawn on (100) plane.

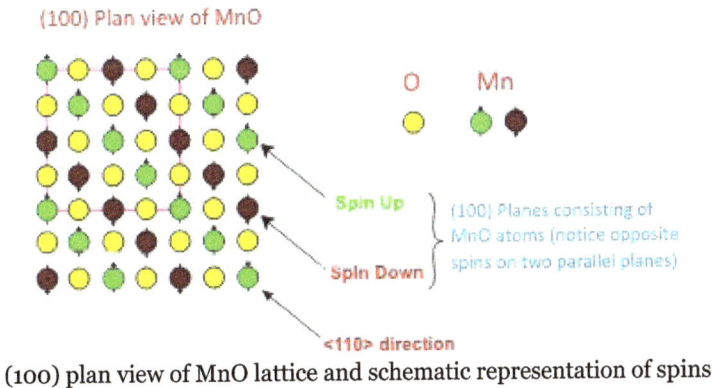

(100) plan view of MnO lattice and schematic representation of spins

As shown below, the susceptibility of an antiferromagnetic material shows a paramagnetic (PM) behavior above (T_N) and between 0 K and (T_N), it shows an antiferromagnetic (AFM) behavior.

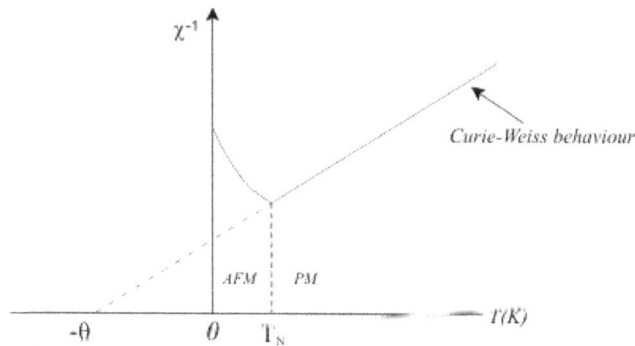

Temperature dependence of susceptibility of an antiferromagnetic material

Antiferromagnetic materials are not very useful as such because they do not show any magnetization whatsoever.

Some of the materials like bismuth ferrite ($BiFeO_3$) have been described as canted antiferromagnets where spin of atoms are not exactly antiparallel but are slightly inclined towards the so-called parallel axis which may give rise to very small magnetization.

Antiferromagnetism

In materials that exhibit antiferromagnetism, the magnetic moments of atoms or molecules, usually related to the spins of electrons, align in a regular pattern with neighboring spins (on different sublattices) pointing in opposite directions. This is, like ferromagnetism and ferrimagnetism, a manifestation of ordered magnetism. Generally, antiferromagnetic order may exist at sufficiently low temperatures, vanishing at and above a certain temperature, the Néel temperature (named after Louis Néel, who had first identified this type of magnetic ordering). Above the Néel temperature, the material is typically paramagnetic.

Measurement

When no external field is applied, the antiferromagnetic structure corresponds to a vanishing total magnetization. In an external magnetic field, a kind of ferrimagnetic behavior may be displayed in the antiferromagnetic phase, with the absolute value of one of the sublattice magnetizations differing from that of the other sublattice, resulting in a nonzero net magnetization. Although the net magnetization should be zero at a temperature of absolute zero, the effect of spin canting often causes a small net magnetization to develop.

The magnetic susceptibility of an antiferromagnetic material typically shows a maximum at the Néel temperature. In contrast, at the transition between the ferromagnetic to the paramagnetic phases the susceptibility will diverge. In the antiferromagnetic case, a divergence is observed in the *staggered susceptibility*.

Various microscopic (exchange) interactions between the magnetic moments or spins may lead to antiferromagnetic structures. In the simplest case, one may consider an Ising model on a bipartite lattice, e.g. the simple cubic lattice, with couplings between spins at nearest neighbor sites. Depending on the sign of that interaction, ferromagnetic or antiferromagnetic order will result. Geometrical frustration or competing ferro- and antiferromagnetic interactions may lead to different and, perhaps, more complicated magnetic structures.

Antiferromagnetic materials occur commonly among transition metal compounds, especially oxides. Examples include hematite, metals such as chromium, alloys such as iron manganese (FeMn), and oxides such as nickel oxide (NiO). There are also numerous examples among high nuclearity metal clusters. Organic molecules can also exhibit antiferromagnetic coupling under rare circumstances, as seen in radicals such as 5-dehydro-m-xylylene.

Antiferromagnets can couple to ferromagnets, for instance, through a mechanism known as exchange bias, in which the ferromagnetic film is either grown upon the antiferromagnet or annealed in an aligning magnetic field, causing the surface atoms of the ferromagnet to align with the surface atoms of the antiferromagnet. This provides the ability to "pin" the orientation of a ferromagnetic film, which provides one of the main uses in so-called spin valves, which are the basis of magnetic sensors including modern hard drive read heads. The temperature at or above which an antiferromagnetic layer loses its ability to "pin" the magnetization direction of an adjacent ferromagnetic layer is called the blocking temperature of that layer and is usually lower than the Néel temperature.

Geometric Frustration

Unlike ferromagnetism, anti-ferromagnetic interactions can lead to multiple optimal states (ground states—states of minimal energy). In one dimension, the anti-ferromagnetic ground state is an alternating series of spins: up, down, up, down, etc. Yet in two dimensions, multiple ground states can occur.

Consider an equilateral triangle with three spins, one on each vertex. If each spin can take on only two values (up or down), there are $2^3 = 8$ possible states of the system, six of which are ground states. The two situations which are not ground states are when all three spins are up or are all down. In any of the other six states, there will be two favorable interactions and one unfavorable one. This illustrates frustration: the inability of the system to find a single ground state. This type of magnetic behavior has been found in minerals that have a crystal stacking structure such as a Kagome lattice or hexagonal lattice.

Other Properties

Synthetic antiferromagnets (often abbreviated by SAF) are artificial antiferromagnets consisting of two or more thin ferromagnetic layers separated by a nonmagnetic layer. Due to dipole coupling of the ferromagnetic layers results in antiparallel alignment of the magnetization of the ferromagnets.

Antiferromagnetism plays a crucial role in giant magnetoresistance, as had been discovered in 1988 by the Nobel prize winners Albert Fert and Peter Grünberg (awarded in 2007) using synthetic antiferromagnets.

There are also examples of disordered materials (such as iron phosphate glasses) that become antiferromagnetic below their Néel temperature. These disordered networks 'frustrate' the antiparallelism of adjacent spins; i.e. it is not possible to construct a network where each spin is surrounded by opposite neighbour spins. It can only be determined that the average correlation of neighbour spins is antiferromagnetic. This type of magnetism is sometimes called *speromagnetism*.

Ferrimagnetic Materials

These are materials which again, like antiferromagnetic materials, show antiparallel alignment of moments at particular atomic sites i.e. magnetic moment of one crystal sub-lattice is anti-parallel to the other. But since most of these materials consist of cations of two or more types, sub-lattices contain two different types of ions with different magnetic moment for two types of atoms and as a result, net magnetization is not equal to zero. The examples of such materials are various kinds of cubic spinel ferrites such as $NiFe_2O_4$, $CoFe_2O_4$, Fe_3O_4 (or $FeO.Fe_2O_3$), $CuFe_2O_4$ etc. Other examples are hexagonal ferrites likes $BaFe_{12}O_{19}$, garnets such as $Y_3Fe_5O_{12}$, represented by a general formula $R_3Fe_5O_{12}$ where R, in addition to yttrium can be one of lanthanide atoms such as lanthanum, cerium, samarium etc.

A schematic representation of this inequality in the neighbouring magnetic moment can be like this:

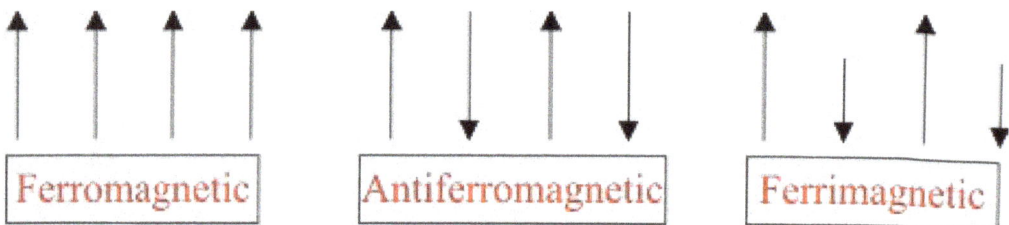

Magnetic moment arrangements in magnetically ordered materials

These materials also follow a temperature dependence of magnetization and susceptibility near Curie transition (actually Néel transition) in a similar manner as shown by the ferromagnetic materials. These materials, like ferromagnetic materials, show significantly large magnetization below the magnetic transition temperature and hence, often the temperature dependent behavior is clubbed with that of ferromagnetic materials.

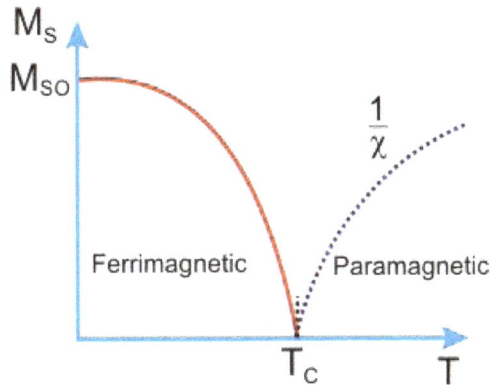

Temperature dependence of magnetization and susceptibility in a ferrimagnetic material

In physics, a ferrimagnetic material is one that has populations of atoms with opposing magnetic moments, as in antiferromagnetism; however, in ferrimagnetic materials, the opposing moments are unequal and a spontaneous magnetization remains. This happens when the populations consist of different materials or ions (such as Fe^{2+} and Fe^{3+}).

Ferrimagnetic ordering

Ferrimagnetism is exhibited by ferrites and magnetic garnets. The oldest known magnetic substance, magnetite (iron(II,III) oxide; Fe_3O_4), is a ferrimagnet; it was originally classified as a ferromagnet before Néel's discovery of ferrimagnetism and antiferromagnetism in 1948.

Known ferrimagnetic materials include YIG (yttrium iron garnet), cubic ferrites composed of iron oxides and other elements such as aluminum, cobalt, nickel, manganese and zinc, hexagonal ferrites such as $PbFe_{12}O_{19}$ and $BaFe_{12}O_{19}$, and pyrrhotite, $Fe_{1-x}S$.

Effects of Temperature

Ferrimagnetic materials are like ferromagnets in that they hold a spontaneous magnetization below the Curie temperature, and show no magnetic order (are paramagnetic) above this temperature. However, there is sometimes a temperature *below* the Curie temperature at which the two opposing moments are equal, resulting in a net magnetic moment of zero; this is called the *magnetization compensation point*. This compensation point is observed easily in garnets and rare earth-transition metal alloys (RE-TM). Furthermore, ferrimagnets may also have an *angular momentum compensation point* at which the net angular momentum vanishes. This compensation point is a crucial point for achieving high speed magnetization reversal in magnetic memory devices.

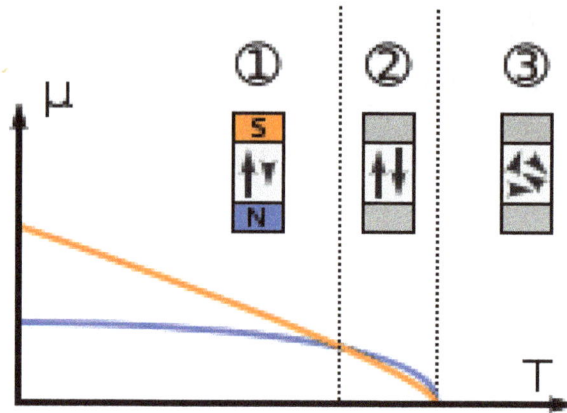

① Below the magnetization compensation point, ferrimagnetic material is magnetic. ② At the compensation point, the magnetic components cancel each other and the total magnetic moment is zero. ③ Above the Curie point, the material loses magnetism.

Properties

Ferrimagnetic materials have high resistivity and have anisotropic properties. The anisotropy is actually induced by an external applied field. When this applied field aligns with the magnetic dipoles it causes a net magnetic dipole moment and causes the magnetic dipoles to precess at a frequency controlled by the applied field, called *Larmor* or *precession frequency*. As a particular example, a microwave signal circularly polarized in the same direction as this precession strongly interacts with the magnetic dipole moments; when it is polarized in the opposite direction the interaction is very low. When the interaction is strong, the microwave signal can pass through the material. This directional property is used in the construction of microwave devices like isolators, circulators and gyrators. Ferrimagnetic materials are also used to produce optical isolators and circulators. Ferrimagnetic minerals in various rock types are used to study ancient geomagnetic properties of Earth and other planets. That field of study is known as paleomagnetism.

Molecular ferrimagnets

Ferrimagnetism can also occur in molecular magnets. A classic example is a dodecanuclear manganese molecule with an effective spin of S = 10 derived from antiferromagnetic interaction on Mn(IV) metal centres with Mn(III) and Mn(II) metal centres.

Comparison between Different Kinds of Magnetism

So far, we have learnt that there are four kinds of magnetism i.e. diamagnetism, paramagnetism, ferromagetism, and ferri- and antiferromagnetism. The field dependent magnetization behavior of these different types is shown below as well the dependence on temperature. Diamagnetism is a basic character present in every material, it is just that this behavior is overshadowed in materials where other effects are present too.

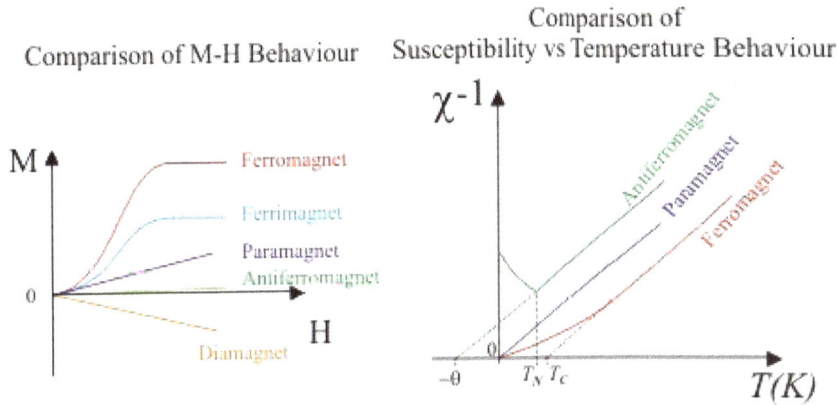

Comparison of magnetic behaviour (magnetization and susceptibility) of various kinds of magnetic materials versus temperature

Practically most important materials are ferromagnets, mostly metals, and ferrimagentic materials, mostly ferrite oxides.

Magnetic Losses and Frequency Dependence

This is important mainly for ferromagnetic and ferrimagnetic materials when they are placed in an alternating magnetic field. Associated with frequency response are magnetic hysteresis losses and frequency response of magnetic permeability.

Magnetic Losses

We imagine a ferromagnetic or ferrimagnetic material with a hysteresis curve is subjected to an alternating magnetic field, $H = H_0 \exp(i\omega t)$. The frequencies are kept such, typically low, such that switching of magnetization of the material keeps pace with the magnetic field switching. In an analogy to the dielectric materials, two loss mechanisms are of importance:

- Eddy current losses induced by the alternating magnetic field which are quite dependent on the resistivity of the material; and

- The domain wall motion requires as well as dissipates some amount of energy resulting in what are called as intrinsic magnetic losses or hysteresis losses.

Both of these effects add up and energy loss is dissipated as heat. While frequency, magnetic field strength and maximum magnetic flux increase both types of losses, increasing the resistivity (ρ) decreases eddy current losses in particular, losses are more when the material is conducting and many ferromagnets tend to be quite conductive.

Hysteresis loss is the energy loss per cycle times number of cycles and is given as

$$P_H = 2.f.H_c.B_r$$

where f is the frequency of applied field, H_c is the coercive field strength and B_r is magnetic remanence.

Eddy current power losses can be expressed as

$$P_e = \frac{\pi d^2}{6\rho}(f.B_{max})$$

The total power loss is the sum of these two losses.

These two equations can help a user to choose a material carefully so that losses can be minimized.

Frequency Response of Permeability

It is given that ferro- and ferri-magnetic materials have permanent moments which constitute regions of aligned moment called as domains with two regions of different regions separated by a domain wall. Now, the ease of domain wall movement can be described in terms of coercivity, i.e. the higher the coercive field, the easier it is for domains walls to move during switching.

So, all we can say at this point is that this domain wall movement is dependent on the frequency of the applied field. In some materials, low frequencies are needed for easy switching while some materials can witness easy switching at GHz or microwave region frequencies making them suitable for microwave devices.

This also means that permeability, μ, in a similar manner as permittivity in dielectrics, can be defined as a complex quantity i.e. μ* = μ' + μ" where μ' is the real part of permeability and μ" is the complex part of the permeability.

Categories of Magnetic Ferrites

Cubic Spinel Ferrites

Cubic spinel ferrites have a formula AB_2O_4 which crystallize with a face centered cubic structure. In these structures, two cations occupy tetrahedral and octahedral sites in an FCC lattice made by O atoms. One unit-cell consists of eight formula units of AB_2O_4 hence containing a total of 32 octahedral interstices with one fourth occupancy and 64 tetrahedral interstices with one eighth occupancy by the cations.

Depending on how these cations are distributed in the interstices, cubic spinel structures can be of two types:

- Normal spinel and

- Inverse spinel

From the magnetism point of interest, the cations occupying tetrahedral sites have their spins oppositely oriented with respect to the cations on octahedral sites (up and down depends upon your frame of reference). Compounds with normal spinel structure are $ZnFe_2O_4$, $MgAl_2O_4$, $CoAl_2O_4$ where A atoms occupy the tetrahedral sites while B atoms occupy the octahedral sites.

In compounds with inverse spinel strcture e.g. Fe_3O_4, $NiFe_2O_4$, half of B cations occupy tetrahedral sites and all A and the remaining 50% B cations occupy octahedral sites.

So, if we take the example of Fe_3O_4 which is actually $Fe^{2+}Fe^{3+}_2O_4$, then the arrangement of spin is like what is shown below.

½ of Fe^{3+} cations (Tetrahedral)	½ of Fe^{3+} and All Fe^{2+} cations (octahedral)

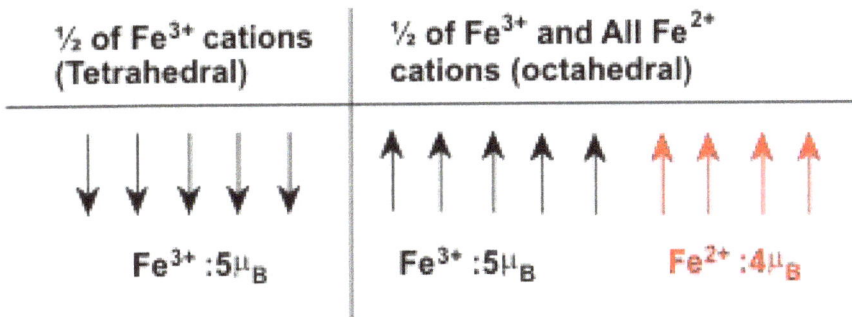

$$\mu_{net} \text{ (per } Fe_3O_4 \text{ formula unit)} = \mu_{oct} - \mu_{tet} = 9\mu_B - 5\mu_B = 4\mu_B$$

Hence the net magnetization of Fe_3O_4 is $4\mu_B$ per formula unit which is quite a large magnetic moment. You can convert this into $A.m^{-1}$ by simply calculating the net moment for the whole unit cell and dividing that by the cell volume.

Similarly you can work out magnetic moments of other ferrites such as $NiFe_2O_4$, $CoFe_2O_4$ etc. This approach is also valid for mixed spinel compounds.

Mixed spinels are quite a nice way of increasing the net magnetic moment. As you can see that since the moments of two sites are antiparallel, reducing the net magnetic moment of one site would actually increase the net moment. So mixing of $NiFe_2O_4$, an inverse spinel, and $ZnFe_2O_4$, a normal spinel, results in maximization of moment up to ~40 mol% $ZnFe_2O_4$.

In general, spinel ferrites show low magnetic anisotropy i.e. dependence of magnetization on crystallographic directions, and are magnetically soft i.e. show low coercive fields. Exceptions could be Co-containing ferrites which are not only strongly magnetically anisotropic but also show large coercive fields strengths. These materials also exhibit a ferromagnetic material like hysteresis loop when placed in a varying magnetic field.

Hexagonal Ferrites

Hexagonal ferrites are based on hexagonal magnetoplubite and are often called M-type ferrites. The model compound of this family is barium ferrite with formula $BaFe_{12}O_{19}$. The large hexagonal unit-cell contains 64 atoms, i.e. two formula units. The structure is basically a mixture of cubic closed packed and hexagonal closed packed layers formed by barium and oxygen ions. Chemical substitution of Ba sites is usually done with Sr atoms while Fe atoms are substituted by Al atoms, based on the size and valence, resulting in a change in the magnetic behavior.

Out of 12 iron atoms of one formula unit, 9 occupy the octahedral sites, two occupy tetrahedral sites and the remaining one is 5-fold coordinated. Out of these, 7 atoms on the octahedral site and 1 with 5-fold coordination have their spins in one direction while spins of the rest of the atoms are oriented oppositely i.e. say 8 atoms with spins up and 4 atoms with spin down. As we saw earlier, each Fe^{3+} ion has spin magnetic moment of $5\mu_B$ simple math gives a net magnetic moment of $20\mu_B$ per formula unit leading to a magnetic moment of $40\mu_B$ per unit cell.

This material has a high degree of magnetic anisotropy and it magnetizes relatively easily along -direction or c-axis of its unit cell. The material is typically categorized as a hard ferrite with coercivity between 50-100 $kA.m^1$ depending upon the microstructure and composition.

Garnets

Garnets are usually known as minerals. In the context of magnetic materials, garnets are represented by a general formula $Y_3Fe_5O_{12}$, containing two magnetic ions, one typically being iron and another being rare earth. Here R, in addition to yttrium can be one of lanthanide atoms such as lanthanum, cerium, samarium etc.

The unit cell of $Y_3Fe_5O_{12}$ is cubic and contains 8 formula units i.e. 160 atoms, quite complex! In garnet ferrites, orbital magnetic contribution of iron atoms is quenched due to shielding from crystal field while lanthanide ions contribute to both orbital and spin magnetic moment, thus contributing more to the total magnetic moment.

In this structure, R atoms are cubic coordinated i.e. 12-fold coordinated, 2 Fe atoms are octahedrally coordinated and the remaining three Fe atoms are tetrahedrally coordinated with antiparallel spin configuration of spins on tetrahedral and octahedral sites while orientation of spins on R-site is parallel to those on octahedral sites. We know that each Fe^{3+} ion contributes $5\mu_B$ which each lanthanide atom, R, contributes a moment of magnitude $\mu_R\mu_B$ where μ_R is the strength of moment of R ion. Hence the total picture looks like the following:

The value of μ_R is 7 for Gd while zero for Y. As we see from the above schematic figure, net magnetic moment would be dominated by rate earth ions when μ_R is greater than $5/3$.

3 R atoms	12-fold coordination		$3\mu_R\mu_B$	
3 Fe atoms	Tetrahedral coordination		$15\mu_B$	$\mu_{net} = (5 - 3\mu_R)\cdot\mu_B$
2 Fe atoms	Octahedral coordination		$10\mu_B$	

This is dependent upon the temperature which governs the coupling between rare earth and Fe ions. Typically the net magnetic moment drops as the temperature increases, especially for strongly magnetic ions like Gd, Tb and Dy. Gd-doped garnet of composition $Y_{1.2}Gd_{1.8}Fe_5O_{12}$ has a rather stable saturation magnetization for a wide temperature range centered around ~50°C.

Garnets can be quite useful materials in microwave applications because of their high electrical resistivity and hence lower losses around microwave frequencies. The material is also easy to synthesize in either of bulk polycrystalline ceramic, single crystal or thin film forms. The structural parameters as well as magnetic properties can be tuned by tailoring the composition of the material.

Properties of Ferrite Ceramics

As discussed above, ferrites can have a broad spectrum of properties depending upon the type of ferrite and compositions. The following table shows some properties of common magnetic ceramics. These values are quite dependent upon the microstructure and processing history of the material. For comparison, values for some common magnetic metals and alloys are also given:

Material	Magnetic permeability, μ_R	Coercive Field $(H_c$, A.m$^-$)	Remanence $(B_s$, T)	Curie temperature $(T_c$, °C)	Resistivity (ρ, Ω,m)
Soft materials					
Fe	150	80	2.16	770	10×10^{-8}
Fe-4%Si	2000	30	1.93	690	60×10^{-8}
Mn-Zn ferrites	500-10,000	5-100	0.35-0.50	90-280	0.01-1
Ni-Zn ferrites	10-2000	15-1600	0.10-0.40	90-500	10^3-10^7
Hard materials					
Medium Carbon steel		4.4	0.9	770	
Isotropic Barium Hexaferrite		200	0.3	450	
Oriented Barium Hexaferrite		320	0.4	450	

Soft Ferrites

Soft ferrites, as we explained earlier, are materials which are easy to magnetize or demagnetize i.e. materials with low coercive field strengths and thus so that they can reverse the direction in alternating fields without dissipating much energy since the area of B-H (or M-H) loop is small.

Typical soft ferrites used in transformer or electromagnetic cores contain nickel, zinc, and/or manganese based ferrites. These materials also have higher resistivities than typical ferromagnetic metals, of the order of 10^{-1} to $10^6 \Omega.m$, which leads to low eddy currents in the core, another source of energy loss.

Because of their comparatively low losses at high frequencies, they are also extensively used in the cores of RF transformers and inductors in applications such as switched-mode power supplies (SMPS). The most common soft ferrites are $Mn_xZn_{(1-x)}Fe_2O_4$, $Ni_x Zn_{(1-x)}Fe_2O_4$. Ferrites of Ni-Zn show higher resistivity than those containing Mn-Zn, and are, therefore, more suitable for frequencies above 1 MHz. Mn-Zn ferrites, in comparison, have higher permeability and saturation induction.

The properties of soft ferrites can be tailored by compositional modifications. For instance, in $Mn_{1-x}Zn_xFe_2O_4$ and $Ni_xZn_{(1-x)}Fe_2O_4$, increasing the Zn content leads to an increase in the magnetic permeability just before the magnetic transition but at the same the magnetic transition temperature also decreases. The increase in magnetic permeability near magnetic transition has been attributed to reduced magnetic anisotropy. The increase in relative permeability is about an order of magnitude in $Mn_{1-x}Zn_xFe_2O_4$ for doping levels up to 50 at % and about 2-3 orders of magnitude in $Ni_{1-x}Zn_xFe_2O_4$ for doping levels up to 70 at %.

Change in grain size also has a profound effect on the relative permeability with permeability increasing with increasing grain size. This is related to the decrease in the grain boundary concentration resulting in less pinning of domain walls by grain boundaries and hence facilitating easy magnetic switching. Electrical resistivity of ferrites is again composition dependent. Electrical conduction in ceramics takes place by hopping of electrons between say two valence states of an ion. In ferrites, d-group elements are susceptible to valence fluctuations. For example Mn-Zn ferrites are more susceptible to valence fluctuations of Mn and Fe as compared to Ni-Zn ferrites. This is also controlled very strongly by processing conditions such as firing temperatures, atmosphere and rate of cooling after sintering.

For example in $Ni_{1-x}Zn_xFe_{2+\alpha}O_{4-\beta}$, if all of the iron is present in 3+ valence state, then $\alpha=0$. Any increase in the iron content i.e. $\alpha > 0$ is compensated by the formation of Fe^{2+} which creates favourable conditions for electron hopping between Fe^{3+} and Fe^{2+} promoting n-type conduction. On the other hand, deficiency of iron i.e. $\alpha < 0$ is usually compensated by oxygen vacancies, resulting in a large increase in the resistivity, about 8 orders of magnitude, as shown below.

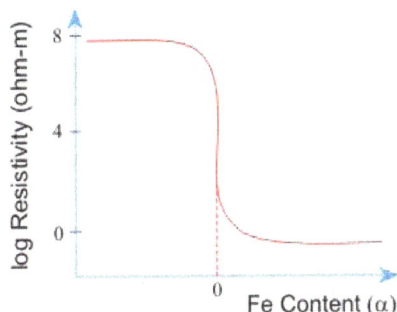

Resistivity variation in $Ni_{1-x}Zn_xFe_{2+\alpha}O_{4-\beta i}$ as a function of Fe content

In addition, the resistivity of nickel ferrites is also increased by addition of small amounts of Cobalt. Reduction of cobalt from Co^{3+} to Co^{2+} state minimizes the reduction of iron to Fe^{2+} state. Since Co ions are sparsely located in the lattice, hopping of electrons between Co^{2+} and Co^{3+} states is minimal. Another method of increasing the resistivity in polycrystalline ferrites can be via grain boundary modification either by preferential oxidation of grain boundaries or by addition of additives like CaO or SiO_2 so that Fe and Ca ions are incorporated into ferrite regions closer to the grain boundaries. Both of these approaches make grain boundaries more resistive than the grains.

Another interest in soft ferrites is the frequency dependence if their properties such as permeability and loss. These properties are very useful for microwave applications, switch mode power supplies, inductors and other high frequency broadband applications.

Properties of Hard Ferrites

Hard ferrites or hard magnets exhibit high coercive fields, typically above 150 kA.m[1] and are often called permanent magnets. This is because these materials are able to withstand any demagnetizing effects that may arise either internally or externally.

In addition to magnetization or remanence (B_r) and coercive field, a permanent magnet is often characterized by product B.H i.e. area under the magnetic hysteresis curve.

Remanence in materials like hexaferrites which are strongly anisotropic can be affected by processing. Since the effect of magnetic anisotropy in a polycrystalline ceramic is small, the material can be synthesized under application of a magnetic field thereby aligning moments along c-axis in the grains giving larger remanence. Similarly, coercivity is a strong function of grain size and it is found for hexaferrites that it is maximum for grain sizes of about 1 μm.

Applications of Magnetic Ceramics

- In electronic inductors, transformers and electromagnets

  Soft ferrites like Mn-Zn and Ni-Zn ferrites are used as core materials in these applications in the frequencies ranging from a 100 kHz to 100 MHz. Typically these ferrites have high electrical resistance which results in very low eddy current losses. Most common radio magnets, including those used in loudspeakers, are ferrite magnets. Ferrite magnets have largely displaced Alnico magnets in these applications.

Ferrites are also used for power transformers which are used to transmit either over a single frequency or within a range such as in ultrasonic generators. For high frequency applications, upto about 5 MHz, Ni-Zn ferrites are useful while for frequencies upto 100 kHz, Mn-Zn ferrites are preferred due to their higher permeabilities.

- Equipment shielding

Here, due to their high impedance to high frequency currents, ferrite components of Ni-Zn and Mn-Zn ferrites are able to prevent high frequency electrical noise due to electromagnetic interference from exiting or entering the equipment.

- Data storage (e.g. magnetic recording tapes and hard disks)

In the magnetic tapes, elongated 0.2-.5 μm long hard magnetic oxide particles of γ - Fe_2O_4 are embedded in nonmagnetic binder. The particles have single domains magnetized along their major axes which are aligned in the plane of the tape. The coercive fields are typically between 50-100 kA.m⁻¹. In magnetic hard-disks, core element is produced by forming several layers of materials (nonmagnetic underlayer, magnetic layer, overcoat, plus layer of lubricants on a nonmagnetic disk substrate). Here, the read/write head is not in direct contact with the hard disk (in contrast to floppy disk) due to an air bearing ($\approx$ 50 nm); air flow is caused by the relative velocity between disk and head. These memories have high storage density of about 10 GB in⁻² and short access time.

Early computer memories stored data in the residual magnetic fields of hard ferrite cores, which were assembled into arrays of core memory. Ferrite powders are used in the coatings of magnetic recording tapes. One such type of material is iron (III) oxide.

- Absorbing materials

In stealth aircrafts, ferrite particles are used as a component of radar-absorbing materials or coatings and in the absorption tile lining in the rooms used for electromagnetic compatibility measurements.

- Microwave applications in the frequency ranges of 1-300 GHz

Materials like Mg-ferrites, Li-doped Ferrites and garnets are used for such applications such as phase shifters, circulators and isolators.

References

- Drakos, Nikos; Moore, Ross; Young, Peter (2002). "Landau diamagnetism". Electrons in a magnetic field. Retrieved 27 November 2012

- Cullity, B. D.; Graham, C. D. (2008). Introduction to Magnetic Materials (2nd ed.). Wiley-IEEE Press. p. 103. ISBN 0-471-47741-9

- A. Herczynski (2013). "Bound charges and currents" (PDF). American Journal of Physics. 81 (3): 202–205. Bibcode:2013AmJPh..81..202H. doi:10.1119/1.4773441

- Yung, K. W.; Landecker, P. B.; Villani, D. D. (1998). "An Analytic Solution for the Force between Two Magnetic Dipoles" (PDF). Magn. Elec. Separation. 9: 39–52. Retrieved November 24, 2012

- Boyer, Timothy H. (1988). "The Force on a Magnetic Dipole". Am. J. Phys. 56 (8): 688–692. Bibcode:1988AmJPh..56..688B. doi:10.1119/1.15501

- Feynman, Richard P.; Leighton, Robert B.; Sands, Matthew (2006). The Feynman Lectures on Physics. 2. ISBN 0-8053-9045-6

- "Search results matching 'magnetic moment'". CODATA internationally recommended values of the Fundamental Physical Constants. National Institute of Standards and Technology. Retrieved 11 May 2012

- Liu, Yuanming; Zhu, Da-Ming; Strayer, Donald M.; Israelsson, Ulf E. (2010). "Magnetic levitation of large water droplets and mice". Advances in Space Research. 45 (1): 208–213. Bibcode:2010AdSpR..45..208L. doi:10.1016/j.asr.2009.08.033

- Buxton, Richard B. (2002). Introduction to functional magnetic resonance imaging. Cambridge University Press. p. 136. ISBN 0-521-58113-3

- Jackson, Roland (21 July 2014). "John Tyndall and the Early History of Diamagnetism". Annals of Science: 4. doi:10.1080/00033790.2014.929743. Retrieved 28 October 2014

- Sessoli, Roberta; Tsai, Hui Lien; Schake, Ann R.; Wang, Sheyi; Vincent, John B.; Folting, Kirsten; Gatteschi, Dante; Christou, George; Hendrickson, David N. (1993). "High-spin molecules: [Mn$_{12}$O$_{12}$(O$_2$CR)$_{16}$(H$_2$O)$_4$]". J. Am. Chem. Soc. 115 (5): 1804–1816. doi:10.1021/ja00058a027

- Furlani, Edward P. (2001). Permanent Magnet and Electromechanical Devices: Materials, Analysis, and Applications. Academic Press. p. 140. ISBN 0-12-269951-3

- Ullah, Zaka; Atiq, Shahid; Naseem, Shahzad (2013). "Influence of Pb doping on structural, electrical and magnetic properties of Sr-hexaferrites". Journal of Alloys and Compounds. 555: 263–267. doi:10.1016/j.jallcom.2012.12.061

- Beatty, Bill (2005). "Neodymium supermagnets: Some demonstrations—Diamagnetic water". Science Hobbyist. Retrieved September 2011

- Steiner, Marcus (2004). Micromagnetism and Electrical Resistance of Ferromagnetic Electrodes for Spin Injection Devices. Cuvillier Verlag. p. 6. ISBN 3-86537-176-0

- F. K. Lotgering, P. H. G. M. Vromans, M. A. H. Huyberts, "Permanent-magnet material obtained by sintering the hexagonal ferrite W=BaFe$_2$Fe$_{16}$O$_{27}$", Journal of Applied Physics, vol. 51, pp. 5913-5918, 1980

- "Fun with diamagnetic levitation". ForceField. 02-12-2008. Archived from the original on February 12, 2008. Retrieved September 2011

A Comprehensive Study of Ferroics and Multiferroics

Ferroic ceramics are materials that exhibit ferromagnetic, ferroelastic or ferroelectric properties. A hysteresis effect is required for such effects to take place. They can be used to make magnetoelectric devices. Electroceramics is best understood in confluence with the major topics listed in the following chapter.

Ferroics

A ferroic material is basically a material which exhibits either ferroelectric or ferromagnetic or ferroelastic ordering, a feature typically demonstrated by the presence of a well defined hysteresis loop when the material is switched electrically, magnetically or mechanically. More recently there has been another ordering mechanism proposed which is called as ferrotoroidic ordering. Magnetoelectric coupling in the materials, on the other hand, is a more general phenomenon irrespective of the state of magnetic and electrical ordering. For example, it could occur in paraelectric ferromagnetic materials or it can be mediated by other parameter such as strain.

Hence, the term multiferroic would mean a material exhibiting two or more of the above ordering mechanisms. More recently, multiferroic materials have become of tremendous interests because of potential device applications. For example, one can have multi-state memory element or sensors which can be operated in multimode or spintronic devices. However, there are challenges in finding a material that would act as a perfect multiferroic. Most multiferroic materials are not naturally occurring and are made in the laboratory. There are problems with respect to their fabricability, while their transition temperatures are often impractical. Despite these challenges, research is on to find a material which would emerge as a potential device material.

Ferroics is the generic name given to the study of ferromagnets, ferroelectrics, and ferroelastics. The basis of this study is to understand the large changes in physical characteristics that occur over a very narrow temperature range. The changes in physical characteristics occur when phase transitions take place around some critical temperature value, normally denoted by T_c. Above this critical temperature, the crystal is in a nonferroic state and does not exhibit the physical characteristic of interest. Upon cooling the material down below T_C it undergoes a spontaneous phase transition. Such a phase transition typically results in only a small deviation from the nonferroic crystal structure, but in altering the shape of the unit cell the point symmetry of the material is reduced. This breaking of symmetry is physically what allows the formation of the ferroic phase.

In ferroelectrics, upon lowering the temperature below T_C, a spontaneous dipole moment is induced along an axis of the unit cell. Although individual dipole moments can sometimes be small, the effect of 10^{24} unit cells gives rise to an electric field that over the bulk substance that is not insignificant. An important point about ferroelectrics is that they cannot exist in a centrosymmetric crystal. A centrosymmetric crystal is one where a lattice point (x, y, z) can be mapped onto a lattice point $(-x, -y, -z)$.

Ferromagnets is a term that most people are familiar with, and, as with ferroelastics, the spontaneous magnetization of a ferromagnet can be attributed to a breaking of point symmetry in switching from the paramagnetic to the ferromagnetic phase. In this case, T_C is normally known as the Curie Temperature.

In ferroelastic crystals, in going from the nonferroic (or prototypic phase) to the ferroic phase, a spontaneous strain is induced. An example of a ferroelastic phase transition is when the crystal structure spontaneously changes from a tetragonal structure (a square prism shape) to a monoclinic structure (a general parallelepiped). Here the shapes of the unit cell before and after the phase transition are different, and hence a strain is induced within the bulk.

In recent years a new class of ferroic materials has been attracting increased interest. These multiferroics exhibit more than one ferroic property simultaneously in a single phase.

Multiferroics

Multiferroics are materials which possess more than one type of primary ferroic ordering in a single phase. The general features are

- Ferroics are materials like ferroelectrics, ferromagnetic or ferroelastics which exhibit a large change in the properties of the materials across a critical temperature and show a characteristic hysteresis loop with two equivalent response states at zero value of stimuli.

- The critical temperature, in general, is also accompanied with a symmetry breaking.

- Typically known orderings are ferroelectric (coupling of charge polarization and electric field), ferromagnetic (coupling of magnetic moment and magnetic field) and ferroelastic (coupling of stress and strain) ordering. Another proposed ordering mechanism is ferrotordoicity which exhibit arrangement of magnetic vortices in an ordered manner, called tordoization.

Figure below explains the various possible scenarios. While there are a large number of magnetically and electrically polarizable materials, there are only a few materials which show ferroelectric and ferromagnetic ordering. Magnetoelectric materials are those materials which are simultaneously electrically and magnetically polarizable, while Multiferroics are strictly those materials which show ferroelectric and ferromagnetic ordering.

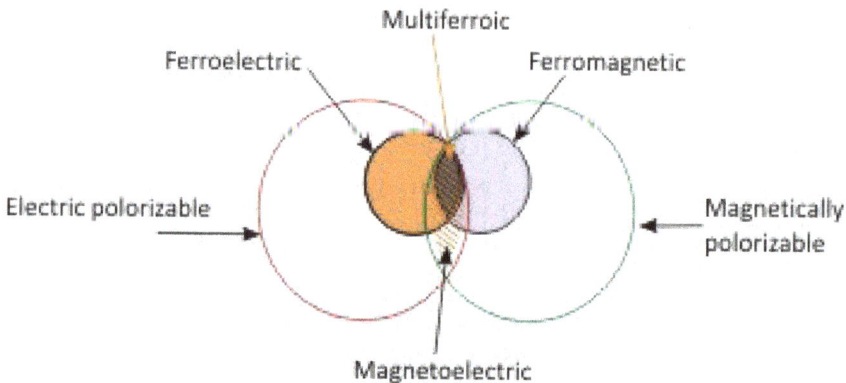

Classification of multiferroic and magnetoelectric materials

While, strictly speaking multiferroism means only for those materials in which there is coupling of more than one order parameter, now a days, researchers have also started including antiferromagnetism as well as ferrimagnetism also with multiferroic materials.

The multiferroic materials are either rare earth manganites or ferrites or transition metal perovskite oxides. The examples are $TbMnO_3$, $TbMn2O_5$, $HoMn2O_5$, $LuFe2O_4$, $BiFeO_3$, $BiMnO_3$ and $YMnO_3$. Some non-oxides are also multiferroics such as $BaNiF_4$ and spinel chalcogenides, e.g. $ZnCr2Se_4$.

Given that the multiferroic materials show more than one ferroic ordering, the envisaged applications are numerous. Some of these applications can be future memory devices with multiple degree of control, sensors and actuators that be controlled by more than one type of stimuli, spintronic devices where spin of electron can be controlled electrically.

Recent reports also classify the multiferroics into Type I and Type II multiferroics. Type I multiferroics are those materials in which the source of ferroelectricity and magnetism is different and the effects are fairly independent of each other, albeit with a small degree of coupling. In contrast, type II materials are those where magnetism causes the existence of ferroelectricity attributed to the strong coupling between two states. However, the magnitude of polarization is these materials remains very small, typically less than 10^{-2} $\mu C/cm^2$.

Magnetoelectric Effect

Magnetoelectric effect was first observed by Rontgen in 1888 and by Pierre Curie in 1894 in two independent studies. Rontgen found that a dielectric when moved in an electric field, became magnetized and conversely it became polarized when moved in a magnetic field. In contrast, Curie pointed out the magnetoelectric effect based on symmetry considerations. The term magnetoelectric was first used by Debye in 1926.

The first material with magneto-electric switching was Cr_2O_3 with small magnitudes of induced polarization and magnetization. Subsequently the research was carried on various materials and it is now established that more than 80 compounds including Ti_2O_3, $GaFeO_3$, boracites, phosphates showed magnetoelectric effect.

The first ever discovered multiferroic material that was simultaneously ferroelectric and ferromagnetic, was nickel iodine boracite, $Ni_3B_7O_{13}I$. Subsequently many studies were made on various boracite compounds. However, most of them had quite complex crystal structures and materials were not very useful from technological viewpoint.

This was followed by studies on mixed perovskites, essentially solid solutions of two perovskite oxide compounds. Russian scientists took the lead in these investigations where they replaced some of the do type cations in the ferroelectric perovskite oxides with magnetic dn type elements in order to induce magnetic ordering. One of first such compounds to be discovered was a solid solution of $Pb(Fe_{2/3}W_{1/3})O_3$ and $Pb(Mg_{1/2}W_{1/2})O_3$. In this compound, ferroelectricity was caused by diamagnetic Mg and W atoms while magnetic ordering is caused by Fe^{3+} ions. Some other candidates were lead based Fe or Co doped tungstates or tantalates which showed ferroelectricity and antiferromagnetic ordering. However, most of these materials had either very low Curie temperatures or Neel temperature which prevented further research on these.

Subsequently, the research focus was on other perovskite materials which are either manganites or ferrites and have been more promising than previously research materials.

Requirements of a Magnetoelectric and Multiferroic Material

There are many material requirements which need to be fulfilled for a material to be called as multiferroic. For instance, for ferroelectricity, a material must be non-centrosymmetric to possess spontaneous electrical polarization and there are only a limited number of point groups (out of 32) which allow an unique polar direction. Similarly, spontaneous magnetic moment is permitted by 31 point groups. Out of these, 13 point groups allow occurrence of both the properties simultaneously. Since this is not a small number; it is probably unlikely that symmetry plays an important role in determining a multiferroic.

Electrically, while a ferroelectric material must be an insulator, it is not a constraint for a ferromagnetic material. For most ferromagnets, electronically speaking, the conductivity is due to high density of states at the Fermi level while the same is not true for ferroelectrics and insulators. However, there are a few magnetic oxides, such as half metallic magnets and ferrimagnetic oxides which show reasonable spontaneous magnetism while simultaneously being semiconducting or insulating.

As far as the chemistry of the material is concerned, most ferroelectrics require ions whose shells are filled and in case of perovskites the B-atom at the centre of BO_6 octahedra must have d0 type electron configuration. In contrast, magnetic systems require d-orbitals to be partially occupied for magnetic ordering to develop. Latter also puts constraints to maintaining the center of symmetry in these systems.

Among type I multiferroics, multiple mechanisms of ferroelectricity have been proposed. For example, in mixed perovskites, it has been suggested that d0 ions being ferroelectrically active shift from the center of O_6 octaehdra while magnetic order is maintained by dn ions. In contrast, in materials like $BiFeO_3$, ferroelectricity is believed to arise due to the ordering of lone pairs of Bi in one direction such as [111]. Another proposed mechanism for ferroelectricity is charge ordering i.e. if after charge ordering has occurred, the sites have different charges and bonds turn out to be of unequal lengths. This is seen in materials like $TbMn_2O_5$. Finally, materials like $YMnO_3$ exhibit geometric ordering due to tilting of rigid MnO_5 polyhedra, resulting in Y and O atoms coming closer to each other forming dipoles.

Another factor that could be analyzed is the size of small cation, especially in the perspective of perovskites. However, upon comparison, one finds that this is not a valid argument as sizes vary considerably for different kinds of compounds.

Another contrast between ferroelectric and ferromagnetically ordered systems is that the way structure is distorted. While ferroelectrics undergo a phase transition as temperature changes, low temperature phase being non-centrosymmetric, ferromagnetic materials show significant Jahn-Teller distortion arising from partially filled d-shells. The latter is almost absent in most ferroelectrics as it has been postulated that Jahn-Teller distorted structure may have less driving force for off-center displacement of B-ions in the octahedra.

Another condition which ferroelectric materials show is that they possess a time reversal symmetry but do not exhibit a space inversion symmetry (i.e. polarization reverses in space). On the other hand, ferromagnetic materials possess space inversion symmetry but do not exhibit time inversion symmetry.

So, in summary, while there is no constraint on various material parameters which prevent materials from being multiferroic i.e. simultaneously ferroelectric and ferromagnetic, a multiferroic does not possess either time reversal or space inversion symmetry.

Magnetoelectric Effect

The magnetoelectric effect (ME) is the phenomenon of inducing magnetic (electric) polarization by applying an external electric (magnetic) field. The effects can be linear or/and non-linear with respect to the external fields. In general, this effect depends on temperature. The effect can be expressed in the following form

$$P_i = \sum \alpha_{ij} H_j + \sum \beta_{ijk} H_j H_k + \dots$$
$$M_i = \sum \alpha_{ij} E_j + \sum \beta_{ijk} E_j E_k + \dots$$

where P is the electric polarization, M the magnetization, E and H the electric and magnetic field, and α and β are the linear and nonlinear ME susceptibilities. The effect can be observed in single phase and composite materials. Some examples of single phase magnetoelectrics are Cr_2O_3, and multiferroics materials which show a coupling between the magnetic and electric order parameters. Composite magnetoelectrics are combinations of magnetostrictive and electrostrictive materials, such as ferromagnetic and piezoelectric materials. The size of the effect depends on the microscopic mechanism. In single phase magnetoelectrics the effect can be due to the coupling of magnetic and electric orders as observed in some multiferroics. In composite materials the effect originates from interface coupling effects, such as strain. Some of the promising applications of the ME effect are sensitive detection of magnetic fields, advanced logic devices and tunable microwave filters.

History of the Magnetoelectric Effect

The magnetoelectric effect was first conjectured by P. Curie in 1894 while the term "magnetoelectric" was coined by P. Debye in 1926. A more rigorous prediction of a linear coupling between electric polarization and magnetization was shortly formulated by L. D. Landau and E. Lifshitz in one book of their famous series on theoretical physics. Only in 1959, I. Dzyaloshinskii, using an elegant symmetry argument, derived the form of a linear magnetoelectric coupling in Cr_2O_3. The experimental confirmation came just few months later when the effect was observed for the first time by D. Astrov. The general excitement which followed the measurement of the linear magnetoelectric effect lead to the organization of the series of MEIPIC (Magnetoelectric Interaction Phenomena in Crystals) conferences. Between the prediction of I. Dzialoshinskii and the MEIPIC first edition (1973), more than 80 linear magnetoelectric compounds were found. Recently, technological and theoretical progress triggered a renaissance of these studies and magnetoelectric effect is still heavily investigated.

Origin of the Magnetoelectric Effect

Single-ion Anisotropy

In crystals, spin-orbit coupling is responsible for single-ion magnetocrystalline anisotropies (provide link) which determine preferential axes for the orientation of the spins (such as easy axes). An external electric field may change the local symmetry seen by magnetic ions and affect both the strength of the anisotropy and the direction of the easy axes. Thus, single-ion anisotropy can couple an external electric field to spins of magnetically ordered compounds.

Symmetric Exchange Striction

The main interaction between spins of transition metal ions in solids is usually provided by superexchange. This interaction depends on details of the crystal structure such as the bond length between magnetic ions and the angle formed by the bonds between magnetic and ligand ions. Symmetric exchange can be both positive and negative and is the main responsible of magnetic ordering. As the strength of symmetric exchange depends on the relative position of the ions, it couples spins to collective lattice distortions, called phonons. Coupling of spins to collective distortion with a net electric dipole can happen if the magnetic order breaks inversion symmetry. Thus, symmetric exchange can provide a handle to control magnetic properties through an external electric field.

Strain Driven Magnetoelectric Heterostructured Effect

Because materials exist that couple strain to electrical polarization (piezoelectrics, electrostrictives, and ferroelectrics) and that couple strain to magnetization (magnetostrictive/magnetoelastic/ferromagnetic materials), it is possible to couple magnetic and electric properties indirectly by creating composites of these materials that are tightly bonded so that strains transfer from one to the other.

Thin film strategy enables achievement of interfacial multiferroic coupling through a mechanical channel in heterostructures consisting of a magnetoelastic and a piezoelectric component. This type of heterostructure is composed of an epitaxial magnetoelastic thin film grown on a piezoelectric substrate. For this system, application of a magnetic field will induce a change in the dimension of the magnetoelastic film. This process, called magnetostriction, will alter residual strain conditions in the magnetoelastic film, which can be transferred through the interface to the piezoelectric substrate. Consequently, a polarization is introduced in the substrate through the piezoelectric process. The overall effect is that the polarization of the ferroelectric substrate is manipulated by an application of a magnetic field, which is the desired magnetoelectric effect (the reverse is also possible). In this case, the interface plays an important role in mediating the responses from one component to another, realizing the magnetoelectric coupling. For an efficient coupling, a high-quality interface with optimal strain state is desired. In light of this interest, advanced deposition techniques have been applied to synthesize these types of thin film heterostructures. Molecular beam epitaxy has been demonstrated to be capable of depositing structures consisting of piezoelectric and magnetostrictive components. Materials systems studied included cobalt ferrite, magnetite, $SrTiO_3$, $BaTiO_3$, PMNT.

Flexomagnetoelectric Effect

Magnetically driven ferroelectricity is also caused by inhomogeneous magnetoelectric interaction. This effect appears due to the coupling between inhomogeneous order parameters. It was also called as flexomagnetoelectric effect. Usually it is describing using the Lifshitz invariant (i.e. single-constant coupling term). It was shown that in general case of cubic hexoctahedral crystal the four phenomenological constants approach is correct. The flexomagnetoelectric effect appears in spiral multiferroics or micromagnetic structures like domain walls and magnetic vortexes. Ferroelectricity developed from micromagnetic structure can appear in any magnetic material even in centrosymmetric one. Building of symmetry classification of domain walls leads to determination of the type of electric polarization rotation in volume of any magnetic domain wall. Existing symmetry classification of magnetic domain walls was applied for predictions of electric polarization spatial distribution in their volumes. The predictions for almost all symmetry groups conform with phenomenology in which inhomogeneous magnetization couples with homogeneous polarization. The total synergy between symmetry and phenomenology theory appears if energy terms with electrical polarization spatial derivatives are taking into account.

Magnetoelectric Coupling

Landau theory describes the magnetoelectric effect in a single phase material through expansion of the free energy expression as

$$F(E,H) = F_0 - P_i^s E_i - M_i^s H_i - \frac{1}{2}\varepsilon_0\varepsilon_{ij}E_iE_j - \frac{1}{2}\mu_0\mu_{ij}H_iH_j - \alpha_{ij}E_iH_j$$

$$+\frac{\beta_{ijk}}{2}E_iH_jH_k + \frac{\gamma_{ijk}}{2}H_iE_jE_k +...$$

(1)

where E and H are the electric and magnetic field respectively. Here ε and μ are the dielectric permittivity and magnetic permeability respectively. The second and the third term in Equation (1) are the temperature dependent electrical polarization, P_i^s, and magnetization, M_i^s. Fourth and fifth terms describe the effect of electrical and magnetic field on the electrical and magnetic behavior respectively, while sixth term consisting of α_{ij} describes linear magnetoelectric coupling. The next two terms consisting of β_{ijk} and γ_{ijk} are third rank tensors and represent higher order coupling coefficients.

Differentiation of Equation (1) with respect to electric and magnetic fields respectively leads to polarization and magnetization which are as follows:

$$P_i = -\frac{\partial F(E,H)}{\partial E_i} = P_i^s + \varepsilon_0\varepsilon_{ij}E_j + \alpha_{ij}H_j + \frac{\beta_{ijk}}{2}H_jH_k +...$$

and

$$M_i = -\frac{\partial F(E,H)}{\partial H_i} = M_i^s + \mu_0\mu_{ij}H_j + \alpha_{ij}E_i + \beta_{ijk}E_iH_j + ... \quad (2)$$

In most cases, we are interested to know about the linear magnetoelectric coefficient, α_{ij}, as magnetoelectric effect is linear in most compounds. This coefficient basically quantifies the dependence of polarization on magnetic field or of magnetization on the electric field. In case of multiferroics, although many linear magnetoelectric effects are expected because these materials often possess large susceptibility and permeability respectively, this is not a necessary condition as some ferroelectrics and ferromagnets do show small dielectric susceptibility and magnetic permeability.

In addition to direct coupling, there may be instances of indirect coupling mediated by strain. This is likely to arise in two phase systems where two components are couple via strain. However, more recently, in cubic $SrMnO_3$ and $EuTiO_3$, strain mediated ME effect is observed in single phase.

Indirect measurements of magnetoelectric coupling include measurement of changes in the magnetization near the magnetic transition temperatures or changes in dielectric constant near the magnetic transition temperature. However, such measurements do not provide any mechanistic insight into the coupling constant. Direct measurements measure magnetic response of material to an applied electric field or electric response to an applied magnetic field.

Type I Multiferroics

There are a few type I single phase multiferroics. As mentioned earlier, Type I multiferroics are the materials which have different sources of ferroelectricity and magnetism with the two effects being quite independent of each other. However, a small degree of coupling cannot be ruled out.

In this section, we will mainly have a look at most studied compounds:

- Bismuth Ferrite (BiFeO_3)

- Bismuth Manganite (BiMnO_3) and

- Hexagonal Manganites

Bismuth Ferrite

Bismuth ferrite ($BiFeO_3$, also commonly referred to as BFO in materials science) is an inorganic chemical compound with perovskite structure and one of the most promising multiferroic materials. The room-temperature phase of $BiFeO_3$ is classed as rhombohedralbelonging to the space group R3c. It is synthesized in bulk and thin film form and both its antiferromagnetic (G type ordering) Néel temperature and ferroelectricCurie temperature are well above room temperature (approximately 653 K and 1100K, respectively). Ferroelectric polarization occurs along the pseudocubic direction ($\langle 111 \rangle_c$) with a magnitude of 90–95 $\mu C/cm^2$.

Sample Preparation

Bismuth ferrite is not a naturally occurring mineral and several synthesis routes to obtain the compound have been developed.

Solid State Synthesis

In the solid state reaction method bismuth oxide (Bi_2O_3) and iron oxide (Fe_2O_3) in a 1:1 mole ratio are mixed with a mortar, or by ball milling and then fired at elevated temperatures. The volatility of bismuth and the relatively stable competing ternary phases $Bi_{25}FeO_{39}$ (sillenite) and $Bi_2Fe_4O_9$ (mullite) makes the solid state synthesis of phase pure and stoichiometric bismuth ferrite challenging. Typically a firing temperature of 800 to 880 Celsius is used for 5 to 60 minutes with rapid subsequent cooling. Excess Bi_2O_3 has also been used a measure to compensate for bismuth volatility and to avoid formation of the $Bi_2Fe_4O_9$ phase.

Single Crystal Growth

Bismuth ferrite melts incongruently, but it can be grown from a bismuth oxide rich flux (e.g. a 4:1:1 mixture of Bi_2O_3, Fe_2O_3 and B_2O_3 at approximately 750-800 Celsius). High quality single crystals have been important for studying the ferroelectric, antiferromagnetic and magnetoelectric properties of bismuth ferrite.

Chemical Routes

Wet chemical synthesis routes based on sol-gel chemistry, modified Pechini routes or hydrothermal synthesis have been used to prepare phase pure $BiFeO_3$. The advantage of the chemical routes is the compositional homogeneity of the precursors and the reduced loss of bismuth due to the much lower temperatures needed. In sol-gel routes, an amorphous precursor is calcined at 300-600 Celsius to remove organic residuals and to promote crystallization of the bismuth ferrite perovskite phase, while the disadvantage is that the resulting powder must be sintered at high temperature to make a dense Polycrystal.

Thin Films

The electric and magnetic properties of high quality epitaxialthin films of bismuth ferrite reported in 2003 revived the scientific interest for bismuth ferrite. Epitaxial thin films have the great advantage that they can be integrated in electronic circuitry. Epitaxial strain induced by singly crystalline substrates with different lattice parameters than bismuth ferrite can be used to modify the crystal structure to monoclinic or tetragonalsymmetry and change the ferroelectric, piezoelectric or magnetic properties. Pulsed laser deposition (PLD) is a very common route to epitaxial $BiFeO_3$ films, and $SrTiO_3$ substrates with $SrRuO_3$ electrodes are typically used. Sputtering, metal organic chemical vapor deposition (MOCVD) and chemical solution deposition are other methods to prepare epitaxial bismuth ferrite thin films. Apart from its magnetic and electric properties bismuth ferrite also possesses photovoltaic properties which is known as ferroelectric photovoltaic (FPV) effect.

Applications

Being a room temperature multiferroic material and due to its Ferroelectric PhotoVoltaic (FPV) effect, bismuth ferrite have several applications in the field of magnetism, spintronics, photovoltaics, etc.

Photovoltaics

In the FPV effect, a photocurrent is generated in a ferroelectric material under illumination and its direction is dependent upon the ferroelectric polarization of that material. So, FPV effect has a promising potential as an alternative to conventional photovoltaic devices. But the main hindrance is that a very small photocurrent is generated in ferroelectric materials like $LiNbO_3$, which is due to its large bandgap and low conductivity. In this direction bismuth ferrite has shown a great potential since a large photocurrent effect and above bandgap voltage is observed in this material under illumination. Most of the works using bismuth ferrite as a photovoltaic material has been reported on its thin film form but in a few reports researchers have formed a bilayer structure with other materials like polymers, graphene and other semiconductors. In a report *p-i-n* heterojunction has been formed with bismuth ferrite nanoparticles along with two oxide based carrier transporting layers. In spite of such efforts the power conversion efficiency obtained from bismuth ferrite is still very low.

Bismuth Manganite

Bismuth manganite is an interesting multiferroic material with a perovskite structure. It is a low temperature ferromagnet and a room temperature ferroelectric. The material shows ferromagnetic ordering below 105 K attributed to the orbital ordering of B-site ions i.e. Mn^{3+} ions and a magnetization of 3.6 μ_B per formula. The material has a perovskite triclinic structure which changes to monoclinic structure at ~450 K and then to a non-ferroelectric orthorhombic phase at ~770K. However, the trouble with this material for device application has been its low resistivity, especially in polycrystalline form. The bulk form of material has been shown to exhibit multiferroic behavior near 80 K and negative magneto-capacitance effect in the vicinity of magnetic transition temperature (T_m) with -0.6% change in the dielectric constant near T_m. The problem which arises with this material is that it requires high pressures in bulk form. In contrast, recent work has shown that it can made resistive in thin film form which can be prepared with much ease.

Hexagonal Manganites (TbMnO$_3$, YMnO$_3$)

Hexagonal manganites are another interesting class of manganites and are depicted by the general formula RMnO$_3$ where R is typically a rare earth ion such as Y and Ho. These materials simultaneously exhibit ferroelectricity and antiferromagentic ordering of magnetic Mn ions. In general, rare earth elements having smaller ionic radii, tend to stabilize hexagonal phase of manganites, RMnO$_3$ (R = Sc, Y, Ho, Er, Tm, Yb, Lu) with space group P6$_3$ cm. In spite of having a chemical formula, ABO$_3$, similar to the perovskites, hexagonal manganites have altogether different crystal and electronic structure. In contrast to the conventional perovskites, hexagonal manganites have their Mn^{3+} ions with 5-fold coordination, located at the center of an MnO$_5$ trigonal bi-prism. R ions, on the other hand, have 7-fold coordination unlike the cubic coordination in perovskites. The MnO$_5$ bi-prisms are two dimensionally arranged in space and are separated by a layer of R^{3+} ions. Figure shows a schematic representation of YMnO$_3$ unit cell showing ionic arrangements within the structure.

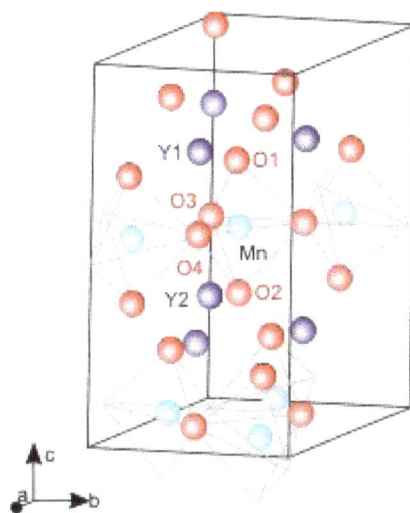

Crystal structure of hexagonal YMnO3

Crystal field level scheme of Mn^{3+} ions in hexagonal RMnO$_3$ is also different from that of Mn^{3+} ions with octahedral coordination. Here, the d- levels are split into two doublets and an upper singlet. As a result, four d-electrons of Mn^{3+} occupy two lowest lying doublets and unlike Mn^{3+} ion in octahedral coordination, there is no degeneracy present. Consequently, Mn^{3+} ions in these compounds are not Jahn-Teller ions.

Hexagonal RMnO$_3$ are found to possess considerably high ferroelectric transition temperature (> 500 K). However, their Neel temperature is far below the room temperature. Table lists the ferroelectric and magnetic transition temperatures, spontaneous polarization (PS) and effective paramagnetic moment μeff of some common $RMnO_3$ along with their structural parameters.

The mechanism of ferroelectricity in these compounds also differs from that of the conventional perovskite oxides. In case of $YMnO_3$, it was observed that off-centering of Mn^{3+} ion from the center of the MnO_5 biprism is very small and cannot be considered to contribute toward ferroelectricity.Apparently it turns out that R ions (Y, here) contributes most toward ferroelectricity by having large R-O dipole moments. However, in reality, ferroelectricity in these materials has different origin and can be considered as accidental by-product. Similar to BO_6 octahedra in perovskite oxides (ABO_3), MnO_5 trigonal biprism in $RMnO_3$, tilts and rotates in order to ensure closest packed structure. Such tilting of MnO_5 trigonal biprism results in loss of inversion symmetry in the structure and brings about ferroelectricity.Since the mechanisms of ferroelectric and magnetic ordering in the above materials are quite different in nature, giant effect of magnetoelectric coupling is understandably not present.

Table: Lattice parameters, Neel temperature (T_N) and ferroelectric Curie (T_C) temperature, effective paramagnetic moment (μe_{ff}) and spontaneous polarization (P_s) of some common hexagonal manganites.

Compound	a(Å)	c(Å)	T_N(K)	T_C(K)	μ_{eff} (in μ_B)	P_s ($\mu C\,cm^{-2}$)
$ScMnO_3$	5.833	11.17	129	-	-	-
$YMnO_3$	6.139	11.39	80	920	89	5.5
$HoMnO_3$	6.142	11.42	76	879	11.1	5.6
$ErMnO_3$	6.112	11.40	80	833	10.5	-
$TmMnO_3$	6.092	11.37	86	>573	8.6	0.1
$YbMnO_3$	6.062	11.36	87	993	6.4	5.5
$LuMnO_3$	6.042	11.37	96	>750	5.2	7.5

Type II Multiferroics

This class of multiferroics is of the materials which show ferroelectricity in their magnetically ordered state and that too of a particular type. Moreover, very strong coupling between ferroelectric and magnetic order parameters has also been observed. In 2003, Kimura et al. reported presence of spontaneous polarization in the magnetized state of the $TbMnO_3$. $TbMnO_3$ has various magnetic structures: it is an incommensurate antiferromagnet between 27 and 42 K and is commensurate antiferromagnet between 7 and 27 K. It is in the commensurate state between 7 and 27 K, the material show ferroelectricity. This discovery was followed by observation of similar effect in $TbMn_2O_5$ by Hur et al. Subsequently variety of other materials have also been investigated such as $Ni_3V_2O_8$, $MnWO_6$ showing this effect. Magnetic spin structure can be either a spiraling cycloid type or a collinear type.

Two Phase Materials

Another method for achieving high degree of magnetoelectric coupling is to mix ferroelectric (e.g. $BaTiO_3$) and ferromagnetic (e.g. $CoFe_2O_4$) materials and utilize the strain between two phases to introduce magneto-electric coupling. Such a coupling requires that two phase have good contact between them i.e. to have an interface through which properties can coupled such as in the form of composites, epitaxial multilayers and laminates. For a few systems, the data is shown in the table below.

Table : Magnetoelectric coupling constant data for selected two-phase magnetoelectric systems

Type of System	Materials	Coupling constant (mV/cm-Oe)
Composite	$BaTiO_3$ and $CoFe_2O_4$	50
Laminated Composite	Terfenol-D in polymer matrix and PZT in polymer matrix	3,000
Laminated	Terfenol-D/PZT	4,800
Laminated	$La0.7Sr_{0.3}MnO_3$ and PZT	60
Laminated	$NiFe_2O_4$ and PZT	1,400
Epitaxial thin film structures	$BaTiO_3$ and $CoFe_2O_4$	--
Epitaxial thin film structures	$BiFeO_3$ and $CoFe_2O_4$	--

In two phase structures, as evident from some of references, one can create large changes in the magnetization owing to strain due to the ferroelectric phase transition of the ferroelectric material during film growth or one can also attempt to alter the magnetic structure by applying a field the piezoelectric material which thereby generates a strain in the magnetic material in the vicinity. Epitaxial growth of layers allows very good interfacial contact between two materials as shown in case of $BaTiO_3$ and $CoFe_2O_4$ which has potential to improve the coupling of parameters.

Multiferroic and magnetoelectric materials are a new class of materials which show interdependence of magnetic and electric properties on each other. Moreover, multiferroics simultaneously exhibit ferroelectric and magnetic ordering in a single phase with some degree of coupling between order parameters. While a multiferroic material has to be a single phase material, magnetoelectric materials can be single phase as well as a mixture of two phases showing interface mediated magnetoelectric coupling. These materials have the potential for a variety of exciting applications such as dual memory devices, spintronic devices, high frequency applications etc. However, the applications are realized yet due to lack of materials and difficulty in achieving the desired effects in the available materials. The single phase materials which have been studied well enough in both bulk and thin film form are $BiFeO_3$, $BiMnO_3$ and hexagonal manganites while two phase mixture studies include $BaTiO_3$ and $CoFe_2O_4$.

References

- Catalan, Gustau; Scott, James F. (26 June 2009), "Physics and Applications of Bismuth Ferrite" (PDF). Advanced Materials. 21 (24): 2463–2485. doi:10.1002/adma.200802849

- Tanygin, B.M. (2011). "On the free energy of the flexomagnetoelectric interactions". Journal of Magnetism and Magnetic Materials. 323 (14): 1899–1902. doi:10.1016/j.jmmm.2011.02.035

- Chu, Ying-Hao; Martin, Lane W.; Holcomb, Mikel B.; Ramesh, Ramamoorthy (2007). "Controlling magnetism with multiferroics" (PDF). Materials Today. 10 (10): 16–23. doi:10.1016/s1369-7021(07)70241-9

- Spaldin, Nicola A.; Cheong, Sang-Wook; Ramesh, Ramamoorthy (1 January 2010). "Multiferroics: Past, present, and future". Physics Today. 63 (10): 38. Bibcode:2010PhT....63j..38S. doi:10.1063/1.3502547. Retrieved 15 February 2012

- Ghosh, Sushmita; Dasgupta, Subrata; Sen, Amarnath; Sekhar Maiti, Himadri (1 May 2005) [14 April 2005]. "Low-Temperature Synthesis of Nanosized Bismuth Ferrite by Soft Chemical Route". Journal of the American Ceramic Society. 88 (5): 1349–1352. doi:10.1111/j.1551-2916.2005.00306

- Chatterjee, S.; Bera, A.; Pal, A.J. (2014). "p–i–n Heterojunctions with BiFeO3 Perovskite Nanoparticles and p- and n-Type Oxides: Photovoltaic Properties". ACS Applied Materials & Interfaces. 6 (22): 20479–20486. doi:10.1021/am506066m

Dipolar Polarization

Dipolar polarization is a polarization that is either inherent to polar molecules (orientation polarization), or can be induced in any molecule in which the asymmetric distortion of the nuclei is possible (distortion polarization). Orientation polarization results from a permanent dipole, e.g., that arising from the 104.45° angle between the asymmetric bonds between oxygen and hydrogen atoms in the water molecule, which retains polarization in the absence of an external electric field. The assembly of these dipoles forms a macroscopic polarization.

When an external electric field is applied, the distance between charges within each permanent dipole, which is related to chemical bonding, remains constant in orientation polarization; however, the direction of polarization itself rotates. This rotation occurs on a timescale that depends on the torque and surrounding local viscosity of the molecules. Because the rotation is not instantaneous, dipolar polarizations lose the response to electric fields at the highest frequencies. A molecule rotates about 1 radian per picosecond in a fluid, thus this loss occurs at about 10^{11} Hz (in the microwave region). The delay of the response to the change of the electric field causes friction and heat.

When an external electric field is applied at infrared frequencies or less, the molecules are bent and stretched by the field and the molecular dipole moment changes. The molecular vibration frequency is roughly the inverse of the time it takes for the molecules to bend, and this distortion polarization disappears above the infrared.

Ionic Polarization

Ionic polarization is polarization caused by relative displacements between positive and negative ions in ionic crystals (for example, NaCl).

If a crystal or molecule consists of atoms of more than one kind, the distribution of charges around an atom in the crystal or molecule leans to positive or negative. As a result, when lattice vibrations or molecular vibrations induce relative displacements of the atoms, the centers of positive and negative charges are also displaced. The locations of these centers are affected by the symmetry of the displacements. When the centers don't correspond, polarizations arise in molecules or crystals. This polarization is called ionic polarization.

Ionic polarization causes the ferroelectric effect as well as dipolar polarization. The ferroelectric transition, which is caused by the lining up of the orientations of permanent dipoles along a particular direction, is called an order-disorder phase transition. The transition caused by ionic polarizations in crystals is called a displacive phase transition.

Ionic Polarization of Cells

Ionic polarization enables the production of energy-rich compounds in cells (the proton pump in mitochondria) and, at the plasma membrane, the establishment of the resting potential, energetically unfavourable transport of ions, and cell-to-cell communication (the Na+/K+-ATPase).

All cells in animal body tissues are electrically polarized – in other words, they maintain a voltage difference across the cell's plasma membrane, known as the membrane potential. This electrical polarization results from a complex interplay between protein structures embedded in the membrane called ion pumps and ion channels.

In neurons, the types of ion channels in the membrane usually vary across different parts of the cell, giving the dendrites, axon, and cell body different electrical properties. As a result, some parts of the membrane of a neuron may be excitable (capable of generating action potentials), whereas others are not.

Dielectric Dispersion

In physics, dielectric dispersion is the dependence of the permittivity of a dielectric material on the frequency of an applied electric field. Because there is a lag between changes in polarization and changes in the electric field, the permittivity of the dielectric is a complicated function of frequency of the electric field. Dielectric dispersion is very important for the applications of dielectric materials and for the analysis of polarization systems.

This is one instance of a general phenomenon known as material dispersion: a frequency-dependent response of a medium for wave propagation.

When the frequency becomes higher:

1. dipolar polarization can no longer follow the oscillations of the electric field in the microwave region around 10^{10} Hz;

2. ionic polarization and molecular distortion polarization can no longer track the electric field past the infrared or far-infrared region around 10^{13} Hz, ;

3. electronic polarization loses its response in the ultraviolet region around 10^{15} Hz.

In the frequency region above ultraviolet, permittivity approaches the constant ε_0 in every substance, where ε_0 is the permittivity of the free space. Because permittivity indicates the strength of the relation between an electric field and polarization, if a polarization process loses its response, permittivity decreases.

Dielectric Relaxation

Dielectric relaxation is the momentary delay (or lag) in the dielectric constant of a material. This is usually caused by the delay in molecular polarization with respect to a changing electric field in a dielectric medium (e.g., inside capacitors or between two large conducting surfaces). Dielectric relaxation in changing electric fields could be considered analogous to hysteresis in changing magnetic fields (for inductors or transformers). Relaxation in general is a delay or lag in the response of a linear system, and therefore dielectric relaxation is measured relative to the expected linear steady state (equilibrium) dielectric values. The time lag between electrical field and polarization implies an irreversible degradation of Gibbs free energy.

In physics, dielectric relaxation refers to the relaxation response of a dielectric medium to an external, oscillating electric field. This relaxation is often described in terms of permittivity as a function of frequency, which can, for ideal systems, be described by the Debye equation. On the other hand, the distortion related to ionic and electronic polarization shows behavior of the resonance or oscillator type. The character of the distortion process depends on the structure, composition, and surroundings of the sample.

Debye Relaxation

Debye relaxation is the dielectric relaxation response of an ideal, noninteracting population of dipoles to an alternating external electric field. It is usually expressed in the complex permittivity ε of a medium as a function of the field's frequency ω:

$$\hat{\varepsilon}(\omega) = \varepsilon_\infty + \frac{\Delta\varepsilon}{1 + i\omega\tau},$$

where ε_∞ is the permittivity at the high frequency limit, $\Delta\varepsilon = \varepsilon_s - \varepsilon_\infty$ where ε_s is the static, low frequency permittivity, and τ is the characteristic relaxation time of the medium. Separating the real and imaginary parts of the complex dielectric permittivity yields:

$$\varepsilon' = \varepsilon_\infty + \frac{(\varepsilon_s - \varepsilon_\infty)}{(1 + \omega^2\tau^2)}$$

$$\varepsilon'' = \frac{(\varepsilon_s - \varepsilon_\infty)\omega\tau}{1 + \omega^2\tau^2}$$

The dielectric loss is also represented by:

$$\tan\delta = \frac{\varepsilon''}{\varepsilon'} = \frac{(\varepsilon_s - \varepsilon_\infty)\omega\tau}{\varepsilon_s + \varepsilon_\infty\omega^2\tau^2}$$

This relaxation model was introduced by and named after the physicist Peter Debye (1913). It is characteristic for dynamic polarization with only one relaxation time.

Variants of the Debye Equation

- Cole–Cole equation

   This equation is used when the dielectric loss peak shows symmetric broadening

- Cole–Davidson equation

  This equation is used when the dielectric loss peak shows asymmetric broadening

- Havriliak–Negami relaxation

  This equation considers both symmetric and asymmetric broadening

- Kohlrausch–Williams–Watts function (Fourier transform of stretched exponential function)

Paraelectricity

Paraelectricity is the ability of many materials (specifically ceramics) to become polarized under an applied electric field. Unlike ferroelectricity, this can happen even if there is no permanent electric dipole that exists in the material, and removal of the fields results in the polarization in the material returning to zero. The mechanisms that cause paraelectric behaviour are the distortion of individual ions (displacement of the electron cloud from the nucleus) and polarization of molecules or combinations of ions or defects.

Paraelectricity can occur in crystal phases where electric dipoles are unaligned and thus have the potential to align in an external electric field and weaken it.

An example of a paraelectric material of high dielectric constant is strontium titanate.

The $LiNbO_3$ crystal is ferroelectric below 1430 K, and above this temperature it transforms into a disordered paraelectric phase. Similarly, other perovskites also exhibit paraelectricity at high temperatures.

Paraelectricity has been explored as a possible refrigeration mechanism; polarizing a paraelectric by applying an electric field under adiabatic process conditions raises the temperature, while removing the field lowers the temperature. A heat pump that operates by polarizing the paraelectric, allowing it to return to ambient temperature (by dissipating the extra heat), bringing it into contact with the object to be cooled, and finally depolarizing it, would result in refrigeration.

Tunability

Tunable dielectrics are insulators whose ability to store electrical charge changes when a voltage is applied.

Generally, strontium titanate ($SrTiO_3$) is used for devices operating at low temperatures, while barium strontium titanate ($Ba_{1-x}Sr_xTiO_3$) substitutes for room temperature devices. Other potential materials include microwave dielectrics and carbon nanotube (CNT) composites.

In 2013 multi-sheet layers of strontium titanate interleaved with single layers of strontium oxide produced a dielectric capable of operating at up to 125 GHz. The material was created via molecular beam epitaxy. The two have mismatched crystal spacing that produces strain within the strontium titanate layer that makes it less stable and tunable.

Systems such as $Ba_{1-x}Sr_xTiO_3$ have a paraelectric–ferroelectric transition just below ambient temperature, providing high tunability. Such films suffer significant losses arising from defects.

Applications

Capacitors

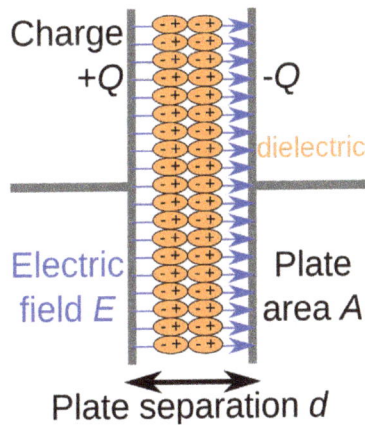

Charge separation in a parallel-plate capacitor causes an internal electric field. A dielectric (orange) reduces the field and increases the capacitance.

Commercially manufactured capacitors typically use a solid dielectric material with high permittivity as the intervening medium between the stored positive and negative charges. This material is often referred to in technical contexts as the *capacitor dielectric*.

The most obvious advantage to using such a dielectric material is that it prevents the conducting plates, on which the charges are stored, from coming into direct electrical contact. More significantly, however, a high permittivity allows a greater stored charge at a given voltage. This can be seen by treating the case of a linear dielectric with permittivity ε and thickness d between two conducting plates with uniform charge density σ_ε. In this case the charge density is given by

$$\sigma_\varepsilon = \varepsilon \frac{V}{d}$$

and the capacitance per unit area by

$$c = \frac{\sigma_\varepsilon}{V} = \frac{\varepsilon}{d}$$

From this, it can easily be seen that a larger ε leads to greater charge stored and thus greater capacitance.

Dielectric materials used for capacitors are also chosen such that they are resistant to ionization. This allows the capacitor to operate at higher voltages before the insulating dielectric ionizes and begins to allow undesirable current.

Dielectric Resonator

A *dielectric resonator oscillator* (DRO) is an electronic component that exhibits resonance of the polarization response for a narrow range of frequencies, generally in the microwave band. It consists of a "puck" of ceramic that has a large dielectric constant and a low dissipation factor. Such resonators are often used to provide a frequency reference in an oscillator circuit. An unshielded dielectric resonator can be used as a dielectric resonator antenna (DRA).

Some Practical Dielectrics

Dielectric materials can be solids, liquids, or gases. In addition, a high vacuum can also be a useful, nearly lossless dielectric even though its relative dielectric constant is only unity.

Solid dielectrics are perhaps the most commonly used dielectrics in electrical engineering, and many solids are very good insulators. Some examples include porcelain, glass, and most plastics. Air, nitrogen and sulfur hexafluoride are the three most commonly used gaseous dielectrics.

- Industrial coatings such as parylene provide a dielectric barrier between the substrate and its environment.

- Mineral oil is used extensively inside electrical transformers as a fluid dielectric and to assist in cooling. Dielectric fluids with higher dielectric constants, such as electrical grade castor oil, are often used in high voltage capacitors to help prevent corona discharge and increase capacitance.

- Because dielectrics resist the flow of electricity, the surface of a dielectric may retain *stranded* excess electrical charges. This may occur accidentally when the dielectric is rubbed (the triboelectric effect). This can be useful, as in a Van de Graaff generator or electrophorus, or it can be potentially destructive as in the case of electrostatic discharge.

- Specially processed dielectrics, called electrets (which should not be confused with ferroelectrics), may retain excess internal charge or "frozen in" polarization. Electrets have a semipermanent electric field, and are the electrostatic equivalent to magnets. Electrets have numerous practical applications in the home and industry.

- Some dielectrics can generate a potential difference when subjected to mechanical stress, or (equivalently) change physical shape if an external voltage is applied across the material. This property is called piezoelectricity. Piezoelectric materials are another class of very useful dielectrics.

- Some ionic crystals and polymer dielectrics exhibit a spontaneous dipole moment, which can be reversed by an externally applied electric field. This behavior is called the ferroelectric effect. These materials are analogous to the way ferromagnetic materials behave within an externally applied magnetic field. Ferroelectric materials often have very high dielectric constants, making them quite useful for capacitors.

Dielectrics are insulating or non-conducting ceramic materials and are used in many applications such as capacitors, memories, sensors and actuators. For the sake of simplicity, we can assume that there is no long range moment of charges. First we will look the simple properties of dielectric materials such as dipole moment, polarization, susceptibility, polarizability and polarization mechanisms. Then we will do analytical treatment of polarizabilities for each of the polarization mechanisms to understand the meaning of these polarizabilities. Subsequently, we will do detailed analysis of dielectric properties for each of the polarization mechanisms under the influence of alternating field, important from the point of understanding the behaviour of these materials in real conditions. Finally, we will look at the breakdown mechanisms which lead to failure of dielectric materials.

Basic Properties: Dielectrics in DC Electric Field

Upon application of a dc or static electric field, there is a long range migration of charges. However, there is a limited movement of charges leading to the formation of charge dipoles and the material, in this state, is considered as polarized. These dipoles are aligned in the direction of the applied field.

The applied field can also align the dipoles that were already present in the material i.e. material containing dipoles without application of the field.

Of course, both these effects may be present in a single material i.e. dipoles can be aligned as well as be induced by the applied field.

The net effect is called Polarization of the material.

Electric Dipole

An electric dipole comprises of two equal and opposite point charges that are separated by a distance δ. The resulting dipole moment, μ is expressed as

$$\mu = q.\delta$$

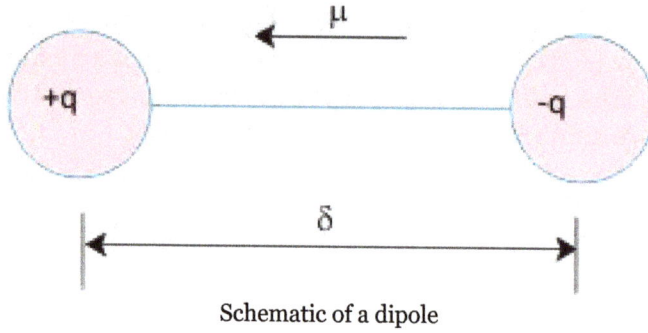

Schematic of a dipole

Dipole moment is a vector and + μ points from −ve to +ve charge. It has units of C.m.

Total dipole moment per unit volume is defined as Polarization, P i.e.

$$P = \frac{\Sigma \mu}{V}$$

Units of P are C.m^{-2} i.e. charge per unit area.

In the most simplistic way when all the dipole are aligned in one direction, P can be written as N.m where N is the number of dipole per unit volume.

You should note that P=0 does not mean that the material does not necessarily have dipole moments rather it is likely that vector summation of all the dipole moment is equal to 0, which is always the case if dipoles were, vectorially speaking, randomly distributed.

We will replace displacement 'δ' by 'd' to avoid duplication by other symbols.

Polarization and Surface Charge

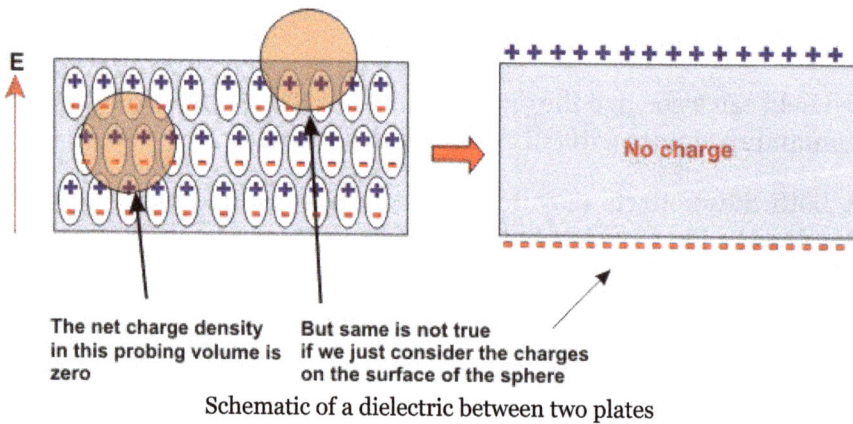

The net charge density in this probing volume is zero   But same is not true if we just consider the charges on the surface of the sphere

Schematic of a dielectric between two plates

Imagine a parallel plate capacitor with homogeneously distributed polarized material between the plates.

Assume that all the dipole moments are aligned in the same direction. Now, if we look at the charge density in a small volume of the material (circle inside the capacitor), then it is clearly zero since both positive and negative charges are equal.

While at the surface, there is a finite charge as shown by the small circle. On both surfaces, the charges move out by a tiny distance, dx, which is nothing but surface polarization charge which can be calculated.

The number of charges, n_c, on the surface area A is equal to the number of dipoles contained within a surface volume (V=A.dx) times the charge of the dipole, q which is nothing but equivalent to one layer of the surface charge.

Assuming homogeneous distribution of the dipoles, polarization can be written as

$$P = \frac{\sum \mu_s}{V_s}$$

where the subscript 's' implies the surface. Hence

$$P = \frac{dx.\sum_s q}{V_s} = \frac{dx.\sum_s q}{dx.A} = \frac{\sum_s q}{A}$$

where $\sum_s$ implies sum over the surface volume. Hence

$$n_c = \sum_s q = P.A$$

So the surface charge density σ is

$$\sigma = \frac{n_c}{A} = |P|$$

σ is a scalar quantity if P.A is scalar.

If polarization is not normal to the surface then

$$\sigma = n.P$$

where n is the outward pointing unit vector normal to the surface of the polarized material.

Dielectric Displacement and Susceptibility

Consider a vacuum plate capacitor configuration as shown below:

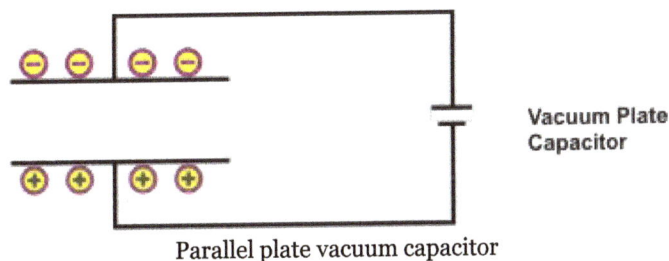

Vacuum Plate
Capacitor

Parallel plate vacuum capacitor

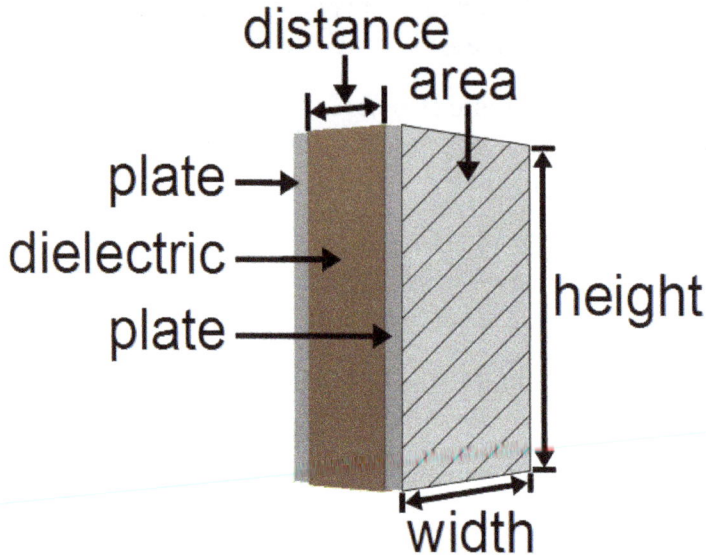

Parallel plate capacitor with a dielectric

For a vacuum capacitor

$$Q = \int I.dt$$

OR

$$Q = CV$$

where capacitance of the vacuum is given as $C_{vac} = (\varepsilon_0 A)(d)$ where ε_0 is the permittivity of free space and is equal to 8.85×10^{-12} F/m.

If one inserts a dielectric between plates, then capacitance gets modified as

$$C = \frac{\varepsilon A}{d} = \frac{\varepsilon_r \varepsilon_0 A}{d}$$

Where εr is dielectric permittivity or more commonly (but not accurately) as relative dielectric constant with value greater than 1.

$$C = \frac{\varepsilon_r \varepsilon_0 A}{d}$$

OR

$$C = \varepsilon_r C_{vac}$$

Thus, inserting a dielectric leads to an increase in the stored charge in the capacitor as shown below.

Basically, ε_r signified some sort of interaction between the material and electromagnetic field.

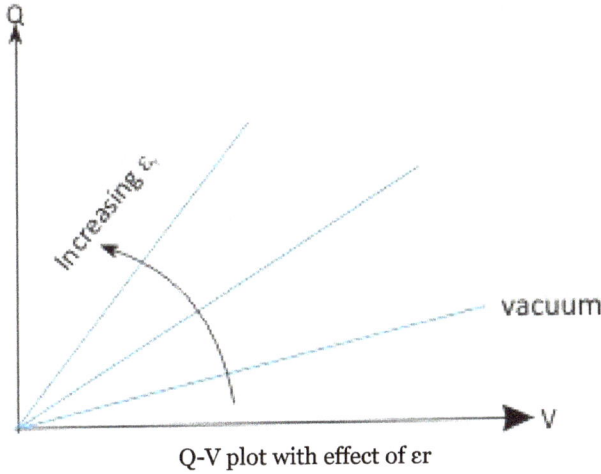

Q-V plot with effect of εr

Polarization Charges

In a parallel plate capacitor without any dielectric, the surface charge in vacuum σ_s is

$$\sigma_s = \left[\frac{Q}{A}\right]_{vac} = \left[\frac{CV}{A}\right]_{vac} = \left[\frac{\varepsilon_o V}{d}\right]_{vac} = \varepsilon_o E$$

where E is the applied field due to the potential V between the plates.

In the presence of a dielectric, the net charge density now becomes

$$\sigma_{net} = \left[\frac{Q}{A}\right]_{dielec} = \frac{\varepsilon_r \varepsilon_o V}{d} = \sigma_s + \sigma_{external}$$

$\sigma_{external}$ is nothing but to due to presence of dielectric and due to polarization of charges and can be written as σ_{pol}. Hence,

$$\sigma_{net} = \sigma_{vac} + \sigma_{pol} = \sigma_{vac} + \rho$$

where ρ is the extra charge resulting from the polarization of the dielectric.

According to the electromagnetic theory, the surface charges on the plates can be defined as dielectric displacement, D, such as

$$D = Q / A$$

OR

$$D = \sigma_{vac} + P$$

OR

$$D = \varepsilon_o E + P$$

The equation shows that the total charge on the plates of a capacitor with dielectric inserted between the plates is now the sum of the surface charge present in a vacuum capacitor ($\varepsilon_0 E$) and extra charge resulting from polarization of the dielectric material, ρ.

Hence, we can now write dielectric displacement as,

$$D = \varepsilon_0 \varepsilon_r E = \varepsilon_0 E + P$$

OR

$$P = \varepsilon_0 E (\varepsilon_r - 1) = \chi . \varepsilon_0 . E$$

where χ is called dielectric susceptibility and is expressed as

$$\chi = \varepsilon_r - 1 = \frac{P}{\varepsilon_0 E}$$

This implies that susceptibility is nothing but the ratio of polarized charge or excess charge to the surface charge in a vacuum capacitor.

The reason for polarization, it can be said that it is basically due to short range movement of masses i.e. electrons, or atoms or molecules under applied electric field. Such a movement is not likely to occur arbitrarily fast, rather it is a function of the frequency of the applied field. This also implies that the dielectric properties are also a function of the frequency. These properties are also a function of material structure and temperature. But for now, we will turn our attention to the understanding of basics of mechanisms of polarization and qualitatively what it means in terms of applied frequency.

Mechanisms of Polarization

Basically, there are four mechanisms of polarization:

Electronic or Atomic Polarization

This involves the separation of the centre of the electron cloud around an atom with respect to the centre of its nucleus under the application of electric field (see (a) in figure below).

Ionic Polarization

This happens in solids with ionic bonding which automatically have dipoles but which get cancelled due to symmetry of the crystals. Here, external field leads to small displacement of ions from their equilibrium positions and hence inducing a net dipole moment (see (b)).

Dipolar or Orientation Polarization

This is primarily due to orientation of molecular dipoles in the direction of applied field which would otherwise be randomly distributed due to thermal randomization (see (c and d)) and finally

Interface or Space Charge Polarization

This involves limited movement of charges resulting in alignment of charge dipoles under applied field. This usually happens at the grain boundaries or any other interface such as electrode-material interface (see (e and f))

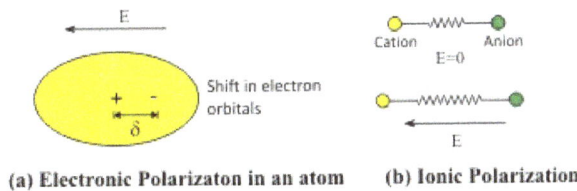

(a) Electronic Polarizaton in an atom   (b) Ionic Polarization

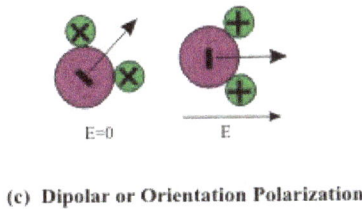

(c) Dipolar or Orientation Polarization

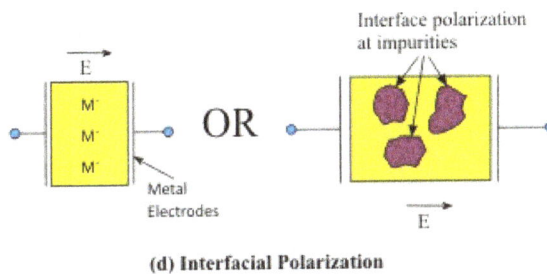

(d) Interfacial Polarization

Schematic of mechanisms of polarization

Atomic polarization is present in all materials by definition and hence any other mechanism would be an addition.

While mathematical treatment of the first three mechanisms is rather straightforward, interface polarization is not easy to quantify.

Qualitatively, we can see that in the above four mechanisms, the masses of the entities to be displaced are different, with mass getting larger from electronic to ionic to dipolar polarization. This has a direct relation with the frequency of the applied field. Intuitively, we can mention that heavier the particular entity, more is the time spent in displacing it. As a result, atomic polarization is the fastest and typically persists at frequencies between $\sim 10^{13} - 10^{15}$ Hz. In contrast, ionic polarization is sluggish and typically occurs at frequencies between $\sim 10^{9} - 10^{13}$ Hz while dipolar polarization involving movement of molecules happens below 10^{9}Hz. Interface or space charge polarization occurs at frequencies below 10 Hz.

For non-magnetic dielectrics, Maxwell's electromagnetic theory predicts that the dielectric constant obtained from the electronic contribution is also related to the index of refraction as $\varepsilon_r = n^2$ which is true at very high frequencies, above $10^{12} - 10^{13}$ Hz. Contribution from any other mechanism will be on top of it. So the total dielectric constant for a material would be $\varepsilon_{r-electronic} \left(= n^2\right) + e_{r-ionic} + \varepsilon_{r-dipolar}$

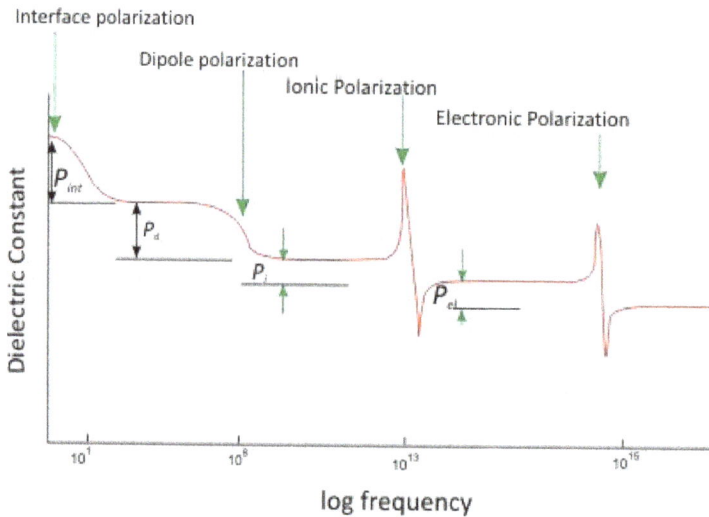

Figure 4.7 Schematic figures between dielectric constant vs frequency showing various mechanisms

The following table shows the values of ε_r and n^2 for a variety of materials and the dominant polarization processes in them:

Material	ε_r	n^2	Dominant mechanisms
C (Diamond)	~5.7	5.85	Electronic
Ge	~16	16.73	Electronic
NaCl	~5.9	2.37	Electronic and Ionic
Water (H_2O)	~80	1.77	Electronic, Ionic and Dipolar

So, you can see that while carbon and germanium being single elemental materials show electronic polarization only and as a result their dielectric constants match well with the values of n^2. However, the same is not the case with NaCl or water which have strong contributions of ionic and ionic and dipolar polarization respectively.

Microscopic Approach

Earlier we saw, polarization in a system with N dipoles per unit volume can be expressed as

$$P = N.\mu$$

i.e.

$$P = N q \delta$$

This gives

$$\chi = \varepsilon_r - 1 = \frac{P}{\varepsilon_o E} = \frac{N q \delta}{\varepsilon_o E}$$

Hence, greater is δ greater is χ and hence larger is ε_r, i.e. more polarizable a medium is, more is its dielectric constant.

Further, polarizability of an ion or atom of type, i, in a dipole is defined as

$$\alpha_i = \frac{\mu_i}{E_{loc}} = \frac{P_i}{N_i E_{loc}}$$

where E_{loc} is the local electric field experienced by an atom or ion or molecule which can be different than the applied field. However, magnitude of local field can be modified quite significantly the polarization of surrounding medium.

As a result P can expressed as

$$P = N_i \, \alpha_i \, E_{loc}$$

Note that above equation relates a macroscopic parameter, P, to the microscopic parameters N, α and E_{loc}.

As a result, in general, susceptibility is

$$\chi = \varepsilon_r - 1 = \frac{N \alpha}{\varepsilon_o}$$

where α is the sum of all types of polarizabilities due to different mechanisms ($\alpha_{electronic} + \alpha_{ionic} + \alpha_{dipolar} + \alpha_{interface}$).

Determination of Local Field

The local field, E_{loc}, experienced by an atom or dipole or molecule usually differs from the applied field, E_{ex} owing to the polarization of the surrounding medium around a dipole or molecule.

While on a macroscopic scale, the overall field in a material is zero due to the condition of electrical neutrality, when we start looking at the scale of individual atoms and molecules, it is not the case. Although this local field which is nothing but the sum of applied field and some other fields and can, in principle, be solved using Poison's equation, coupling charge density and potential at each location, the method is far from being simple. Instead, we use a simple approach, by using Lorentz model where we divide the field into different components.

The understanding of this field can be obtained from the figure shown below which shows a sphere of dielectric material, say containing about 10 atoms, removed. The local field at the center of this sphere is composed of two macroscopic fields and two microscopic fields.

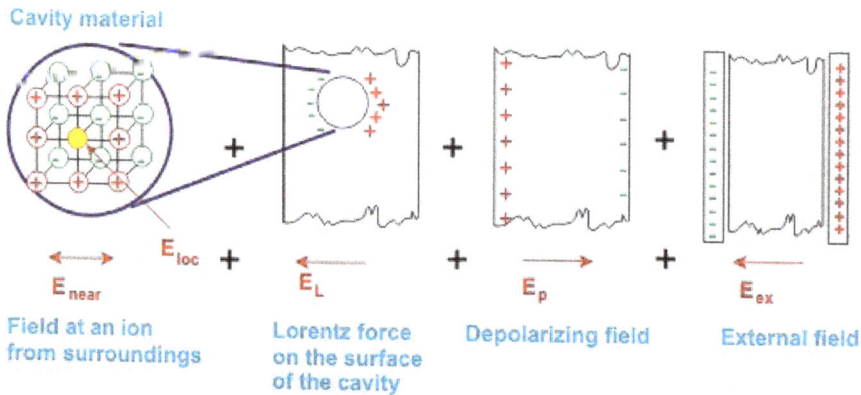

Schematic representation of various electric fields when a small cavity is created in a dielectric

The microscopic fields are:

- E_{center}, field at the center of the sphere due to contributions of ions surrounding it with-in the sphere.

- E_L, Lorentz field at the center of the spherical cavity due to charges on the surface of the cavity

(Note the difference between the two: one is talking of a sphere while it is composed of material and another is talking of a material surrounding the spherical cavity from which material has been removed.)

The macroscopic fields considering the dielectric as a continuum are:

- E_{ex}, field due to applied voltage across the capacitor

- E_d, the depolarizing field due to dielectric polarization

Hence, E_{loc} is written as

$$E_{loc} = \left(E_{ex} - E_d\right) + E_L + E_{center}$$

Now, we know that

$$E_{ex} - E_d = E - \frac{P}{\varepsilon_0} \text{ if } \varepsilon_r = 1$$

Lorentz showed that for isotropic crystals, $E_{centre} = 0$

E_L can be worked out for a cavity as

$$E_L = \frac{P}{3\varepsilon_0}$$

Hence the local field is given as

$$E_{l\alpha} = E + \frac{P}{3\varepsilon_0}$$

By substituting $P = \varepsilon_0(\varepsilon_r - 1)E$ in above equation

$$E_{loc} = \frac{E}{3}(\varepsilon_r + 2)$$

Now combining above equations, we get

$$P = N_i\alpha_i \frac{E}{3}(\varepsilon_r + 2)$$

Combining above equations, we get

$$(\varepsilon_r - 1)\varepsilon_o = N_i\alpha_i \frac{E}{3}(\varepsilon_r + 2)$$

OR

$$\frac{N_i\alpha_i}{3\varepsilon_o} = \frac{\varepsilon_r - 1}{\varepsilon_r + 2}$$

The above equation is called Clausius-Mossotti relationship and is valid only for linear dielectrics. The equation is related to a macroscopic quantity on RHS, i.e. ε_r with a microscopic quantity, α.

As a result, if one knows the dielectric constant by means to measurement, this form can be used to calculate the polarizabilities of a variety of materials to quite an accurate estimate.

Polarizability, α, is the sum of polarizabilities from all the contributing mechanisms i.e. $\alpha_{electronic}$, α_{ionic}, $\alpha_{dipolar}$ and $\alpha_{interfacial}$.

Special case:

For gases

$$\varepsilon_r - 1 << 1$$

And hence at low pressures

$$\varepsilon_r + 2 \approx 3$$

Thus

$$\frac{N\alpha}{\varepsilon_0} = \varepsilon_r - 1 = \chi$$

Analysis of the Lorentz Field

Here, we are interested in calculating the field from the free ends of dipoles i.e. Lorentz field E_l, lined along the cavity wall in the direction of applied field, as shown below. This charge density arises from the bound charges and is determined by the normal component of polarization/dielectric displacement vector P and is written as

$$\vec{P}.n.dA = P \cos \theta.dA$$

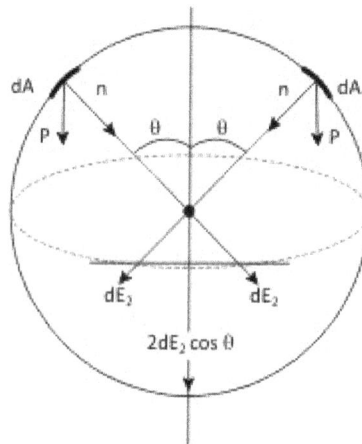

Schematic of field components for a spherical cavity

Now, since each element dA contributes to the field, according to Coulomb's law, the radial field intensity is

$$dE_l = \frac{P \cos \theta}{\varepsilon_0 4\pi r^2} dA$$

Each dA's angular position is between θ and $\theta + d\theta$ and for each dA element, there is another dA element on the other side of the sphere which produces same but opposite horizontal field component.

Horizontal components cancel each other and vertical components $dE_2 \cos\theta$ survives

So the total field intensity is

$$E_l = \oint_{sphere} \frac{P\cos^2\theta}{\varepsilon_o 4\pi r^2} dA$$

The field intensity is parallel to the applied field and actually strengthens it. Now we can also rewrite dA as

$$dA = 2\pi r \sin\theta d\theta$$

So

$$E_l = \int_0^\pi \frac{P\cos^2\theta}{\varepsilon_0 4\pi r^2} 2\pi r^2 \sin\theta d\theta = \frac{P}{3\varepsilon_0}$$

Effect of Alternating Field on the Behavior of a Dielectric Material

Here we will examine the behavior of real and ideal dielectric materials under the influence of an alternating electric field, giving an account of the real situations to which dielectric materials are subjected.

Behavior of an Ideal Dielectric

While most of the above discussion has been for d.c. or static fields, in most practical applications, dielectrics are used under alternating fields. Hence, it is imperative to work out their characteristics in alternating fields.

Let us apply a sinusoidal field

$$V = V_o . \exp(i\omega t) \tag{1}$$

This leads to the development of a charging current, I_c, due to a change in the charge with time which is

$$I_c = \frac{dQ}{dt} = C.\frac{dV}{dt} = i\omega CV \tag{2}$$

$$= \omega C.V_o . \exp(i\omega t).\exp\left(i\frac{\pi}{2}\right)$$

$$= \omega C.V_o . \exp\left[i\left(\omega t + \frac{\pi}{2}\right)\right] \tag{3}$$

The term $+ \pi/2$ implies that the current leads the voltage by 90° in a perfect dielectric. This current voltage relationship can also be understood from a phasor diagram as shown below.

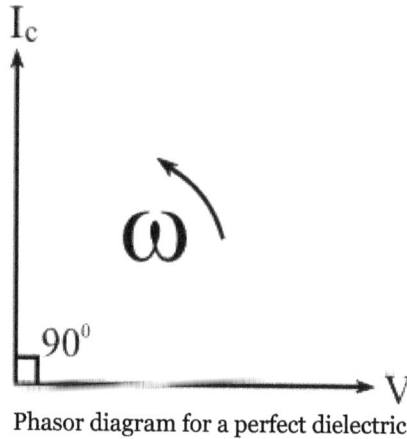

Phasor diagram for a perfect dielectric

Power Dissipation in an Ideal Dielectric

Since the instantaneous power drawn by the dielectric from the voltage sources is $I_c V$, the time-average power dissipated, P_{avj}, in a dielectric is given as

$$P_{avj} = \frac{1}{\Gamma} \int_0^\Gamma I_c . V . dt$$

(4)

where Γ is the time period is given as $(2\pi/\omega)$.

For an ideal dielectric with no loss of charging current, the power dissipated inside the dielectric must be zero, i.e.

$$P_{avj} = \frac{1}{\Gamma} \int_0^\Gamma (-\omega C) . V_0^2 . \sin(\omega t) \cos(\omega t) dt$$

(5)

What it means is that during the first cycle, the capacitor charges completely and during the other cycle it completely discharges without any loss of charge. It is like a mass perfectly oscillating under gravity on a perfect spring.

Behavior of Real Dielectrics

However, in real dielectrics, the charging current is also accompanied by a loss current, associated with storage of electric charge by the dipoles. There are two sources of this loss current:

- Long range migration of charges due to ohmic conduction and is frequency independent i.e. d.c. in nature, and

- Dissipation of energy due to dipole rotation or oscillation as there always is certain inertial to their movement due their mass. This contribution is time-dependent i.e. frequency dependent.

Since, both of these current are in phase with the applied field, the loss current, Il, can be written as

$$I_l = \left(G(\omega)_{ac} + G_{dc} \right).V \tag{6}$$

where G depicts the conductance in mho or Siemens which is nothing but the inverse of resistance.

Hence, the total current, I_L, is the sum of charging and loss current i.e

$$I_T = I_c + I_l = \left(i\omega C + G(\omega)_{ac} + G_{dc}.V \right) \tag{7}$$

Hence, the current in a real dielectric is a complex quantity that leads voltage by angle $90° - \delta$ where δ is defined as the loss angle or dissipation factor. It is made up of two components that are $90°$ out of phase with respect to each other and have to be added vectorially as shown below. Naturally, δ will be zero if there was no loss current present.

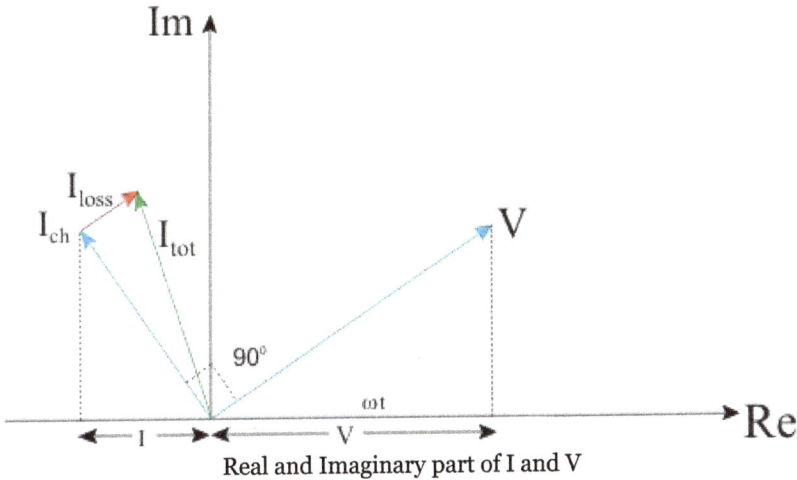

Real and Imaginary part of I and V

When the field applied is static or d.c. i.e. $\omega = 0$ then

$$I_T = I_l = G_{dc}.V$$

where $G_{dc} = 1/R$ and R is the ohmic resistance.

An alternative way to express the real dielectrics passing both charging and loss current is to use complex permittivity.

$$\varepsilon = \varepsilon' - i\varepsilon'' \tag{8}$$

and

$$k^* = \frac{\varepsilon}{\varepsilon_o} = k' - ik'' \tag{9}$$

Total current in a dielectric can now be expressed in terms of a single material parameter, k, since capacitance and charge are

$$C = k^* C_0 \text{ and } Q = CV = kC_o V \tag{10}$$

Using equation (1), the total current can be expressed as

$$I_c = I_T - I(\omega)$$

$$= I_T - I(\omega = 0) = \frac{dQ}{dt}$$

$$I_T - I(\omega = 0) = \frac{dQ}{dt} = k^*.C_o.i\omega V$$

$$= (k' - ik'').C_o.i\omega V \tag{11}$$

i.e.

$$I_T = i\omega C_o k' V + \omega k'' C_o V + G_{dc} V \tag{12}$$

Here

$$I(\omega = 0) = G_{dc}.V \tag{13}$$

The first term in (12) is the out-of-phase charging current term and last two terms are in-phase loss current terms.

Comparing equation (12) with 7) yields overall conductance, G, as

$$G = G_{dc} + G(\omega)_{ac} \tag{14}$$

where $G(\omega)_{ac} = \omega k'' C_0 V$

Loss tangent, tan δ, is expressed as

$$\tan \delta = \frac{I_L}{I_c} = \left(\frac{G_{dc} + \omega k'' C_o}{\omega k' C_o} \right) \tag{15}$$

If $G_{dc} << \omega k'' C_0$ then

$$\tan \delta = k'' / k' \tag{16}$$

The values of dielectric constant and loss tangent for selected materials are shown below.

Material	Dielectric Constant ($\varepsilon_r^{'}$)	Loss Tangent ($\tan\delta^{*}10^{-4}$)
Alumina	10	5-20
SiO_2	3.8	2
$BaTiO_3$	500	150
Nylon	3.1	10
Polycarbonate	~3	10
PVC	3	160

Power Dissipation in a Real Dielectric

Hence, a.c. conductivity can be expressed as

$$\sigma_{ac} = \sigma_{dc} + \omega k^{"} = \omega\varepsilon_o k^{'} \tan\delta \tag{17}$$

If σ_{dc} is very small, then

$$\sigma_{ac} = \omega\varepsilon_o k^{'} \tan\delta \tag{18}$$

Now, the time average power loss can be expressed as

$$P_{av} = \frac{1}{\Gamma}\int_0^{\Gamma} I_{loss} V dt$$

$$= \frac{1}{\Gamma}\int_0^{\Gamma}\left(\omega k^{"} C_o + G_{dc}\right) V_0^2 \cos^2\left(\omega t\right) dt \tag{19}$$

If $G_{dc} \ll \omega k^{"} C_0$, then

$$P_{av} = \frac{1}{2}\left(\omega k^{"} C_o\right).V_0^2 = \frac{1}{2} G_{ac} V_0^2 \tag{20}$$

OR

$$P_{av} = \frac{1}{2}\left(\omega k^{'} \tan\delta C_o\right).V_0^2 = \frac{1}{2} V_0^2 .\omega.C.\tan\delta \tag{21}$$

Hence, if $C_0 = \left(\varepsilon_0 A / d\right)$, $C_0 = \left(V_0 / d\right)$ and $V = A.d$, the dissipated power density in a dielectric would be

$$\frac{P_{av}}{V} = \frac{1}{2}\omega.\varepsilon_o k^{'}.\tan\delta.E_0^2$$

One can now see that for static or d.c. fields, $\omega = 0$, i.e.

$$G_{ac} = G_{dc} = \frac{1}{R} \qquad (22)$$

and hence,

$$P_{av} = I^2 R \qquad (23)$$

which is the standard equation for power dissipation in a material under dc fields.

Dipolar Relaxation

- Relaxation processes occur in many ceramics and ionic solids, especially glasses and these materials applying damped forced oscillator approach does not work in the low frequency region.

- In such cases, the structure of the solid plays an important role because movement of charges may have to occur over many atomic distances and can be classified as diffusional in nature. These processes can be strongly temperature dependent in nature.

- As a result, it may take considerable amount of time in the distribution of charges.

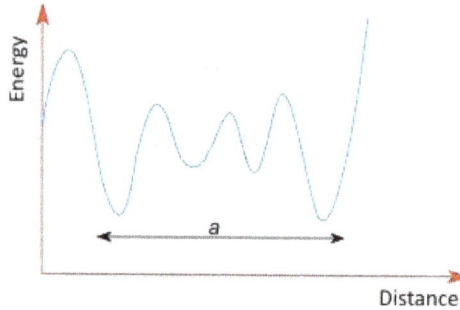

Potential energy distribution of ionic sites in a glass

- For instance, in an ionic solid, application of a field leads to first almost instantaneous development of ionic and electronic polarization followed by slow development of dipolar polarization, P_d to a saturation polarization, P_s.

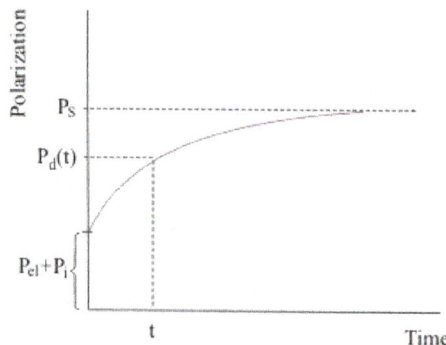

Dependence of dipolar polarization on time

- We can approximately express the rate of change of polarization as

$$\frac{dP}{dt} = \frac{1}{\tau}(P_s - P_d(t)) \qquad (24)$$

- where $1/\tau$ is the proportionality constant. The above equation is also called Relaxation Equation. This equation can be derived using a simple bi-stable model.

Bi-stable Model for Dipolar Relaxation

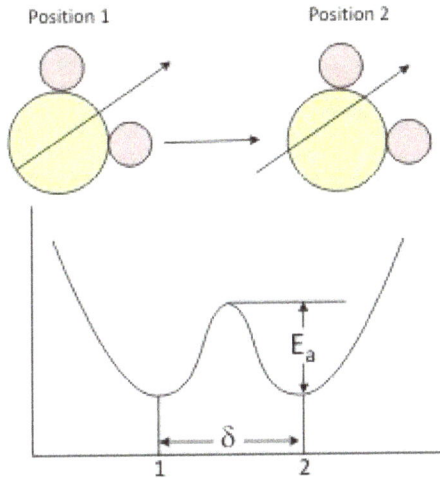

Schematic of a polar molecule going one state to another and resulting energy well diagram

In a solid when field is applied to a polar material, the ions hop from ionic position to another as shown above in the figure, e.g. Na^+ movement in glasses. We consider a bistable dipole model. As the cation moves from left to right, there is a change in the coordinates. At any temperature above 0K, there is random oscillation of cation between these sites.

The probability of jump between sites is given as

$$F = Ae^{-Ea/kT} \qquad (25)$$

Upon application of field, the wells tilt in the direction of applied field, resulting in unequal probability in two directions resulting a net flow of dipoles

$$F_{12} = F\left(1 - \frac{\mu E}{kT}\right)$$

$$F_{21} = F\left(1 - \frac{\mu E}{kT}\right) \qquad (26)$$

where E is the applied field.

Under the application of an ac field, the change in the number of dipoles at site 1 = outflow to site 2 – inflow to site 1 i.e.

$$\frac{dN_1}{dt} = -N_1 F_{12} + N_1 F_{21}$$

(27)

Where $N_1 + N_2 = N = $ constant

Further, since

$$\frac{dN_1}{dt} + \frac{dN_2}{dt} = 0$$

$$\Rightarrow \frac{dN_1}{dt} = -\frac{dN_2}{dt}$$

(28)

$$\Rightarrow \frac{d(N_1 + N_2)}{dt} = 2\frac{dN_1}{dt} = -2\frac{dN_1}{dt}$$

Replace dN_1/dt in the above equation, and we get

$$\frac{1}{2}\frac{d(N_1 + N_2)}{dt} = -N_1 F_{12} + N_2 F_{21}$$

$$\Rightarrow \frac{1}{2}\frac{d(N_1 + N_2)}{dt} = -N_1 F\left(1 - \frac{\mu E}{kT}\right) + N_2 F\left(1 - \frac{\mu E}{kT}\right)$$

$$\frac{1}{2}\frac{d(N_1 + N_2)}{dt} = -F(N_1 - N_2) + F(N_1 - N_2)\frac{\mu E}{kT}$$

(29)

Now, polarization P can be expressed as the product of the net ions moved and the dipole moment i.e.

$$P = (N_1 - N_2)\mu$$

$$\frac{1}{2\mu}\frac{dP}{dt} = -\frac{F}{\mu}\bar{P} + \frac{FN\mu E}{kT}$$

$$\frac{1}{2F}\frac{dP}{dt} + P = \frac{N\mu^2 E}{kT}$$

(30)

This is a relaxation equation with characteristic relaxation time $\tau = 1/2F$ where F is the jump probability in s^{-1}.

So, the above equation can be written as in the form of "Dipolar Polarization" by substituting P with $P_d(t)$ and μ^2/kT by dipolar polarizability leading to

$$\tau\frac{dP_d}{dt} + P_d(t) = N\alpha_s E = P_s \quad OR$$

$$\frac{dP_d}{dt} = \frac{1}{\tau}(P_s - P_d(t))$$

(31)

Solution of Relaxation Equations

The initial and final conditions are

At $t = 0$, $P_d = 0$ and at some reasonably large t, $P = P_s$

Integrating within these limits yield

$$P_d = P_s(1 - \exp(-\tfrac{t}{\tau}))$$ (32)

where τ is defined as relaxation time.

Application of alternating fields to this polarization is not as simple as in the previous analysis.

This is because the saturation polarization, P_s, is dependent on the instantaneous value of field and thus will be time-dependent and the local field is a function of both position and time.

In any case, if we assume that polarizing field is expressed as $E^* = E_0 \exp(\text{1wt})$ then P_s can be expressed as

$$P_s = (\varepsilon_{rs}' - \varepsilon_{r\infty}')\varepsilon_0 E^*$$ (33)

Here, we can define ε_{rs}' as static dielectric constant or dielectric constant at very low frequencies and $\varepsilon_{r\infty}'$ is dielectric constant at high frequencies covering electronic and ionic polarization. This kind of makes sense because dipolar polarization occurs between these two ends of the frequency.

Now, substitute E^* and P_s into (31)

$$\frac{dP_d}{dt} = \frac{1}{\tau}[((\varepsilon_{rs}' - \varepsilon_{r\infty}')\varepsilon_0 E_0 \exp(i\omega t) - P_d(t)]$$ (34)

The integration of this equation yields us

$$P_d = C.\exp\left(-\frac{t}{\tau}\right) + \frac{\varepsilon_{rs}' - \varepsilon_{r\infty}'}{1 + i\omega t}.\varepsilon_0 E^*$$ (35)

Here the first term is the transient time dependent term depicting decay of the d.c. charges on the capacitor and second term is the a.c. behavior of the polarization under an alternating field.

Since, electronic and ionic polarization are approximately frequency independent in this regime of frequencies, we can write

$$\varepsilon_{rs}' - \varepsilon_{r\infty}' = \frac{P_d}{\varepsilon_0 E^*}$$ (36)

Now ignoring the transient time dependent term of (27) and substituting (28) into (26), we get

$$\varepsilon'_{rs} = \varepsilon'_{r\infty} + \frac{\varepsilon'_{rs} - \varepsilon'_{r\infty}}{1 + i\omega t}$$

(37)

We also know that $\varepsilon_r^* = \varepsilon_r' - i\varepsilon_r''$, hence, now we can separate the real and imaginary parts as

$$\varepsilon_r' - i\varepsilon_r'' = \varepsilon'_{r\infty} + \frac{\varepsilon'_{rs} - \varepsilon''_{r\infty}}{1 + i\omega t}$$

OR

$$\varepsilon_r' = \varepsilon'_{r\infty} + \frac{\varepsilon'_{rs} - \varepsilon''_{r\infty}}{1 + \omega^2 \tau^2}$$

(38)

$$\varepsilon_r'' = \frac{\omega\tau}{1 + \omega^2\tau^2}(\varepsilon_r' - \varepsilon'_{r\infty})$$

(39)

and the loss tangent is

$$\tan\delta = \frac{\varepsilon_r''}{\varepsilon_r'} = \frac{(\varepsilon'_{rs} - \varepsilon'_{r\infty})\omega\tau}{\varepsilon'_{rs} + \varepsilon'_{r\infty}\omega^2\tau^2}$$

(40)

The above equations are called Debye equations.

The graphical representation of these is shown below:

Here, the relaxation frequency ω_r is defined as inverse of the relaxation time i.e. $1/\tau$.

The equation (31) and the above figure suggest that ε_r' is independent of frequency at values corresponding to the sum of three polarizations i.e. $P_d + P_i + P_e$. As the applied field frequency approaches value $\omega = 1/\tau$, ε_r' passes an inflection and then drops off to $\varepsilon_{r\infty}'$ which is dependent only on $(P_i + P_e)$.

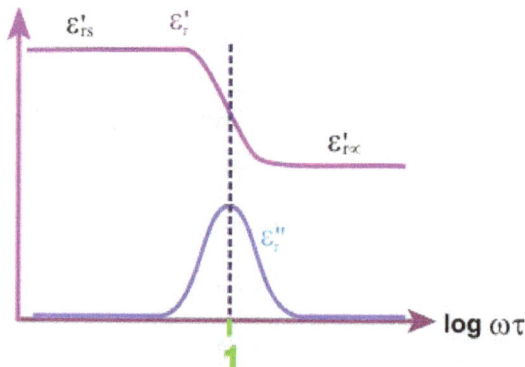

Graphical representation of Debye equations

At $\omega = 1/\tau$, the oscillating charges are coupled with the applied field and absorb maximum energy as depicted by the peak in ε_r'' magnitude of ε_r'' peak being $(\varepsilon_{rs}'' - \varepsilon_{r\infty}')/2$ depending upon the number of oscillating charges and distance of motion.

It is also seen that $\tan\delta$ is also goes than a maximum, but at higher frequencies with $\omega = (\varepsilon_{rs}'/\varepsilon_{r\infty}')^{1/2}/\tau$.

Now since, we have understood that polarization develops by temperature dependent diffusional processes which also give rise to the *d.c.* conductivity, temperature dependence of the relaxation time, τ can be expressed as

$$\tau = \tau_0 . \exp\left(\frac{Q_a}{kT} \right) \tag{41}$$

where Q_a is the activation energy for dipole relaxation and τ_0 is the intrinsic relaxation time.

Combining Debye equations with (32) we can obtain the frequencies ($= \omega\tau$) at which maxima for ε_r'' as well as $\tan\delta$ occur and you will that these frequencies for the maxima change with temperature because τ is temperature dependent. The plots below show this trend for $Li_2O.2SiO_2$ glass ceramic.

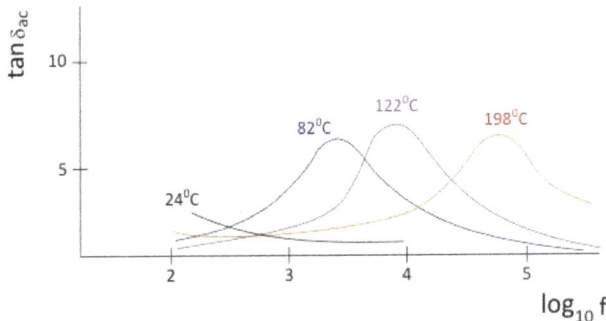

Shift of $\tan\delta$ peak in $Li_2O.2SiO_2$ glass due to increasing temperature

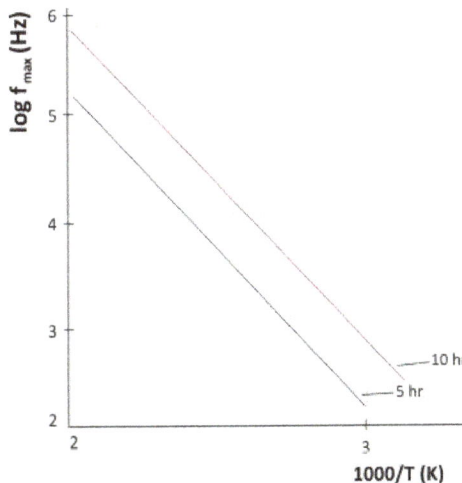

Temperature dependence of the frequency maxima vs time for $Li_2O.2SiO_2$ glass at 500°C

Complete Picture of Frequency Dependence of the Dielectric Constant

So, now we can plot the contributions to the dielectric constant from all the mechanisms.

In case of an ideal dielectric material exhibiting all four basic mechanisms, we would expect the following curve.

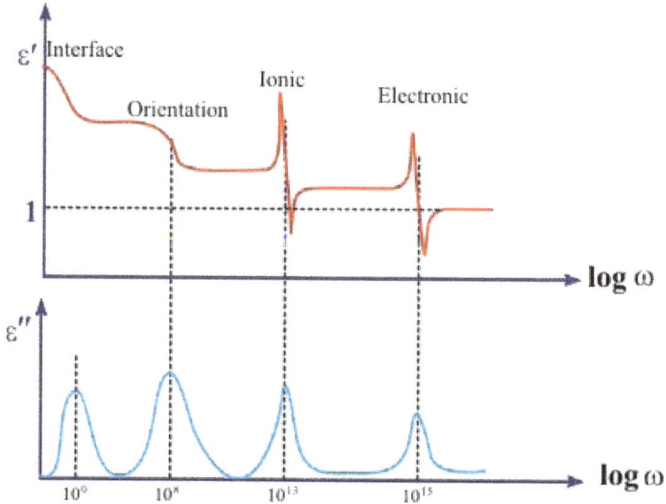

Complete plot of frequency dependence of dielectric constant and loss
(note that frequency is not to scale)

Although the above plot represents an ideal material, yet the plot gives an idea of what you might expect when you measure dielectric constant of a material as a function of frequency.

Although the real plots may look quite different, you can expect a correlation between the real and imaginary part of the curve i.e. we can still clearly see the absorption peak.

As, mentioned earlier, for a non-magnetic dielectric solid, Maxwell's electromagnetic equations predicts that ε_r be equal to n^2.

Circuit Representation of a Dielectric and Impedance Analysis

Data analysis after dielectric characterization often requires modeling of dielectrics which is helped by their representation as equivalent electrical circuits. Incidentally, a perfect dielectric material can be modeled by an equivalent RC parallel circuit as shown below.

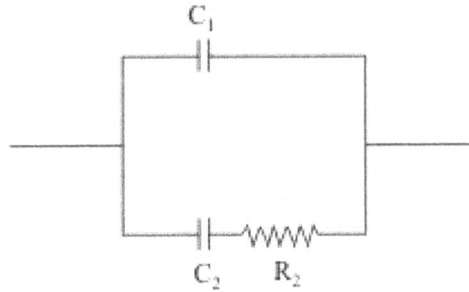

Equivalent circuit model of a dielectric

Let us consider the admittance (Y) of the material representing the above circuit which is inverse of the impedance (Z), and is expressed as

$$Y = \frac{1}{Z} = \frac{1}{Z_1} + \frac{1}{Z_2}$$

(42)

where $Z_1 = \frac{1}{i\omega C_1}$ and $Z_2 = \frac{1}{i\omega C_2} + R_2$

(43)

Hence admittance now is

$$Y = i\omega C_1 + \frac{1}{R_2 + \frac{1}{i\omega C_2}}$$

(44)

Now if we consider time constant for the segment $R_2 C_2$ as $T_2 = R_2 C_2$, then admittance can be written as

$$Y = i\omega C_1 + \frac{C_2}{\tau_2 + \frac{1}{i\omega}}$$

$$= \frac{\omega^2 \tau_2 C_2}{1 + \omega^2 \tau_2^2} + i\omega \left(C_1 + \frac{C_2}{1 + \omega^2 \tau_2^2} \right)$$

(45)

Now, since admittance can be related to the dielectric constant as

$$Y = \frac{1}{Z} = \left(\varepsilon'' + i.\varepsilon' \right) \frac{\omega C_o}{\varepsilon_o} = \left(\varepsilon_r'' + i.\varepsilon_r' \right) \omega C_o$$

(46)

This gives

$$\varepsilon_r'' = \frac{C_2}{C_o} \cdot \frac{\omega \tau_2}{1 + \omega^2 \tau_2^2} \qquad \varepsilon_r' = \frac{C_1}{C_o} + \frac{C_2}{C_o} \cdot \frac{1}{1 + \omega^2 \tau_2^2}$$

and

(47)

which have the same form as Debye equations with $C_2 = \left(\varepsilon_s - \varepsilon_\infty \right) C_o$ and $C_1 = \varepsilon_\infty$

In 1941, K.S. Cole and R.H. Cole explained the behavior of dielectrics in alternating fields in which they plotted the electrical response of dielectric materials to the alternating fields as a function of frequency.

By this technique, they were able to identify and relate the observed relaxation effect with the atomic and microstructural features of the materials.

However, Cole and Cole used modified Debye equation which is

$$\frac{\varepsilon'_r - \varepsilon'_{r\infty}}{\varepsilon'_{rs} - \varepsilon'_{r\infty}} = \frac{1}{(1+i\omega t)^{1-\alpha}}$$

(48)

where α is a parameter which describes the distribution of relaxation times in the material.

For an ideal dielectric with well defined relaxation time i.e. $\alpha = 0$, the system would be represented by the Debye equations i.e. equations (37-39).

As a result, if we plot $\varepsilon''_r(\omega)$ vs $\varepsilon'_r(\omega)$ i.e. imaginary vs real part of the dielectric constant on a complex plane, we get a semi-circle.

Dielectric Breakdown

Every material is bound to fail or breakdown under certain conditions. Basically, in case of a dielectric it means short circuiting across the dielectric.

Technically speaking, dielectric breakdown occurs the electron density in the conduction band becomes very high during the application of an electric field such that conductivity increases rapidly resulting in a permanent damage to material. However, it is easier to measure and talk in terms of the electric current which is anyway a representation of electron density.

The most critical parameter is the field strength E in the dielectric. If it exceeds a critical limit, breakdown occurs. The (DC) current vs. field strength characteristic of a dielectric therefore may look like this:

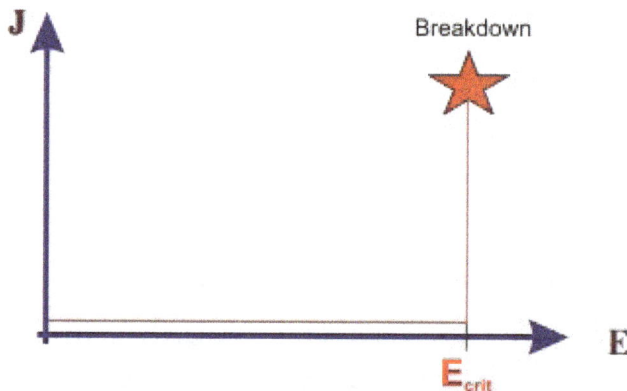

Dielectric breakdown I-V Plot

After reaching a critical field or breakdown field, E_c, a sudden increase in the current may, within a few seconds or even quicker, completely destroy the dielectric resulting in something like a 'burnt' material. However, E_c is not a well defined material property, it depends on many parameters such as material thickness (bulk or thin film), temperature, atmosphere, level of porosity, crystalline anisotropy, level of crystalinity and composition.

While electric field plays an important role, dielectric may also break in a gradual time dependent manner and in such cases we would rather call it as 'failure'. In such situation, the field may be well below the nominal breakdown or critical field and material is destroyed in long time. In such cases, normally the breakdown field also decreases with time.

In such cases, the breakdown may not be sudden, rather a leakage current develops which increases over time, and it may develop until it suddenly increases leading to complete failure. You can do this measurement rather easily by letting a small current pass through the samples and then monitor the voltage needed as a function of time. You will notice that the voltage needed to pass this current reduces as time progresses indicating that materials is getting leakier.

A typical voltage-time curve may then look like this:

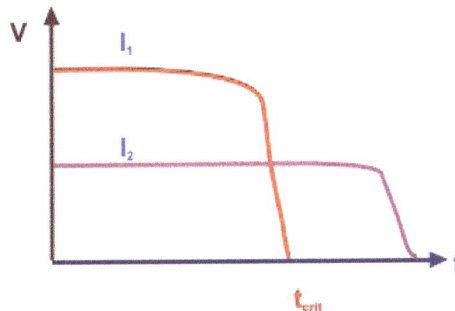

Time dependent failure for a dielectric

The values of breakdown fields for some materials are given below:

Material	Critical Field Strength [kV/cm]
Oil	200
Glass, ceramics	200-400
Mica	200-700
Oiled paper	1800
Polymers	50-900
Al_2O_3 film (100 nm)	16,000
Al_2O_3 ceramic	200-300
$BaTiO_3$ (bulk single crystal)	300
$BaTiO_3$ (Polycrystalline ceramic)	120
SiO_2 (in Integrated circuits)	> 10,000

Example:

For example, in thin film memory devices, SiO_2 is used as a gate dielectric and has a thickness of a few nanometers say 5 nm. The voltages at which these devices operate are about 5 V which translates into a field of 10 MV/cm which is a very large field when compared to break down fields of most of the bulk materials. This explains the importace of material form on the breakdown field.

Basic Mechanisms of Breakdown

- Intrinsic breakdown

- Thermal breakdown

- Avalanche breakdown

Intrinsic Breakdown

- This mechanism is based on lattice ionization and subsequent increase in the electron temperature.

- Actual breakdown field is larger than critical breakdown field, E_c, needed to cause a critical breakdown electron temperature, T_c.

- This mechanism is field dependent and here appliedfield determines the electron temperature to reach critical level for breakdown

- The break time is very short, smaller than ms, suggesting that the process is electronic is nature.

- It is independent of sample or its geometry or waveform type.

- It is a purely material dependent process.

Thermal Breakdown

- It occurs due to heat dissipation in the sample due to current flowing through defective parts of the sample which in turn further increase the ionic defect concentration leading to subsequent increase in the conductivity and then failure.

- It is a very common process in most of the bulk materials.

- It depends on the speed of application of field.

- It is observed between room temperature and 300°C.

- Ambient temperature determines the electron temperature and not the electric field strength.

- Rate of application of field is an important factor.

- The process can be quite slow, from minutes to ms, and is dependent on sample geometry

- The shorter the pulse time is, the higher is the breakdown voltage.

Avalanche Breakdown

- Large electric field in the samples lead to energetic electrons which can further lead to a multiplication process *i.e.* a few electrons knock out more and more electrons leading to a large increase in the conductivity.

- There is a gradual build up of charge rather than sudden change in conductivity even through charge build up can be quite fast.

- Quite often it occurs in thin films.

- It occurs at low temperature and in short time.

Other or Pseudo breakdown mechanisms are

- Dielectric discharge

- Electrochemical and/or mechanical breakdown

Dielectric discharge

- In small pores which are always present in sintered dielectric ceramics, the field strength is higher than the average field and as a result, microscopic arc discharge in the pores may be initiated.

- Electrons and ions from the discharge bombard the inner surface and erode it. As the pores grow, the current in the arc increases leading to an increased sample temperature eventually leading to failure.

Electrochemical breakdown

- This occurs due to transport of conducting material due to the presence of local electrochemical current paths or defect into the interior of the dielectric leading to overall increase in the sample conductivity and then failure.

- It is assisted by suitable atmospheric conditions such as humidity and pH.

Nonlinear Dielectrics

So far, we have discussed linear dielectrics whose dielectric constant increases linearly with the applied field accompanied by an increase in the polarization depending upon the presence of polarization mechanisms in the materials.

In addition, there are a few special classes of dielectric materials which show large dielectric constants, non-zero polarization in the absence of electric field and non-linearity in the dielectric constant. These also show extraordinary special effects such as

- Coupling of strain and electric field (piezoelectric ceramics),

- Temperature dependence of the polarization (pyroelectric ceramics) and

- Presence of large polarization in absence of electric field *i.e.* spontaneous polarization (ferroelectric ceramics).

Most of these materials happen to be oxides and as you can very well understand now, these properties will be greatly affected by the defect chemistry and process variables.

The presence of these features makes these materials extremely useful for a variety of applications such as sensors, actuators, transducers, temperature detectors, imaging, permanent data storage etc.

We will discuss origin of these properties with a crystallographic and thermodynamic framework and associated mathematical representations along with a few examples of materials and devices.

Classification Based on Crystal Classes

Out of a total of 32 crystal point groups, 21 are non-centrosymmetric *i.e.* crystals not having a center of symmetry.

The term centrosymmetric refers to a space group which contains an inversion center as one of its symmetry elements *i.e.* for every point (x, y, z) in the unit cell, there is an indistinguishable point (-x, -y, -z).

Crystal class	Centro symmetric Point groups		Noncentrosymmetric Point groups				
		Polar	Non-polar				
Cubic	m3	m3m	none		432	$\bar{3}$ m	23
Tetragonal	4 or m	4 or mmm	4	4mm	$\bar{4}$	$\bar{4}$ 2 m	22
Orthorhombic	mmm		mm2		222		
Hexagonal	6 or m	6 or mmm	6	6mm	$\bar{6}$	$\bar{6}$ 2m	622
Trigonal	$\bar{3}$	$\bar{3}$ m	3	3m	32		
Monoclinic	2 or m		2	m	none		

Triclinic	$\bar{1}$	1	none
Total Number	11 groups	10 groups	11 groups

Out of these 21 point groups, except group 432, crystals containing all other point groups exhibit piezoelectric effect *i.e.* upon application of an electric field, they exhibit strain or upon application of an external stress, charges develop on the faces of crystal resulting in an induced electric field.

Out of these 20 non-centrosymmetric point groups, 10 belong to polar crystals *i.e.* crystals which possess a unique polar axis, an axis showing different properties at the two ends.

These crystals can be spontaneously polarized and polarization can be compensated through external or internal conductivity or twinning or domain formation.

Spontaneous polarization depends upon the temperature. Consequently, if a change in temperature is imposed, an electric charge is developed on the faces of the crystal perpendicular to the polar axis. This is called pyroelectric effect. All 10 classes of polar crystals are pyroelectric.

In some of these polar non-centrosymmetric crystals, the polarization along the polar axis can be reversed by reversing the polarity of electric field. Such crystals are called ferroelectric i.e. these are spontaneously polarized materials with reversible polarization.

So by default, all ferroelectric materials are simultaneously pyroelectric and piezoelectric. Similarly, all pyroelectric materials are by default piezoelectric but not all of them are ferroelectric.

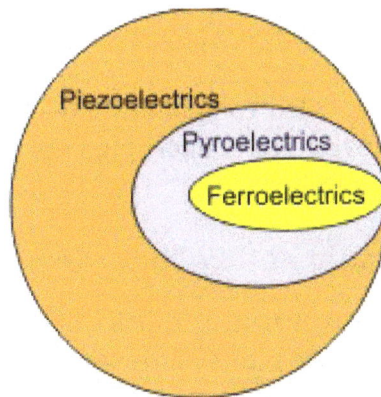

Classificatin of piezo-, pyro- and ferro-electrics

Ferroelectric Ceramics

Ferroelectric ceramics is a special group of minerals that have ferroelectric properties: the strong dependence of the dielectric constant of temperature, electrical field, the presence of hysteresis and others.

The first widespread ferroelectric ceramics material, which had ferroelectric properties not only in the form of a single crystal, but in the polycrystalline state, i.e. in the form of ceramic barium titanate was $BaO \cdot TiO_2$, which is important now. Add to it some m-Liv not significantly change its properties. A significant nonlinearity of capacitance capacitor having ferroelectric ceramics materials, so-called varikondy, types of VC-1 VC-2, VC-3 and others.

Ferroelectric effect was first observed in Rochelle salt $KNaC_4H_4O_6.4H_2O$ by Czech scientist Roger Valasek in 1921. Afterwards, for years the discovery did not raise much excitement possibly due to the world war. However, after a few decades, re-newed technological interest led to much more extensive studies and a better un-derstanding.

In the subsequent years, many materials were discovered and were of technological interest as they were employed into a variety of applications. Among various categories of ferroelectric materials following stand out either due to interest in the structure or in the properties:

- Tri-glycine sulfate and isomorphous materials

- Pottasium dihyrogen phosphates and isomorphous materials

- Barium titanate and other perovskite structured compounds such as $KNbO_3$, $PbTiO_3$ etc

- Complex oxides such as Aurrivillious compounds

- Rochelle Salt and similar compounds

- Ferroelectric Sulfates

Among the above, perovskite related compounds have been studied most, primarily due to their exciting properties and reasonably high transition temperatures, making them attractive for various applications.

Permanent Dipole Moment and Polarization

These materials consist of net permanent dipole moment *i.e.* finite vector sum of dipole moment even in the absence of electric field. This requires the material to be non-centrosymmetric whereas dipole moment would be forced to be zero in a centrosymmteric material due to symmetry considerations.

On top of this, there must be a spontaneous polarization as well which means the cen-ters of positive and negative charges in a crystal would never be the same.

The following is the figure showing the crystal structure of a perovskite structured ma-terial such as $BaTiO_3$. Here in cubic structure (A), the dipole is zero while in tetragonal form (B), the dipole moment is finite. This displaced position of the central atom is the energetically stable position i.e. the free energy is minimum.

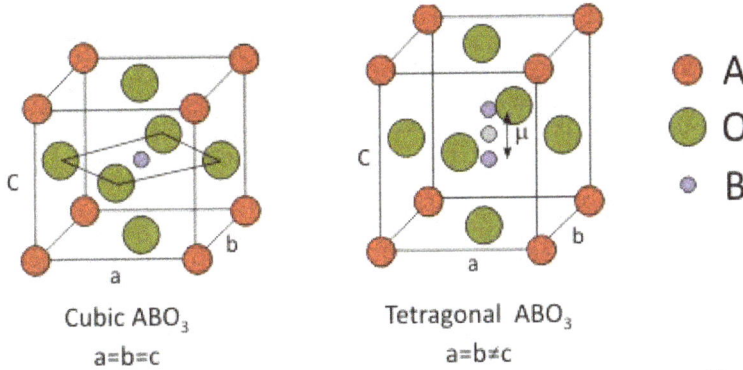

Cubic ABO$_3$ Tetragonal ABO$_3$

a=b=c   a=b≠c

Cubic and tetragonal perovskite structures with dipole moment in latter represented by an arrow due to movement of central B ion up or down along c-axis

When this kind of ferroelectric material is switched i.e. subjected to a bipolar electric field, it exhibits a polarization *vs* electric field plot as shown below. You can see that there is finite polarization in the absence of electric field i.e. two equal and opposite values, +P$_r$ or -P$_r$ also called remnant polarization. These two values can be connected with the position of a central atom in the octahedra in the figure above. So, when you apply field in the direction, the central atom moves up with respect to oxygen octahedral and when you change the polarity, it comes down in the opposite direction, the extent of displacement being similar.

The plot shows a hysteresis which is important for applications such as memory devices.

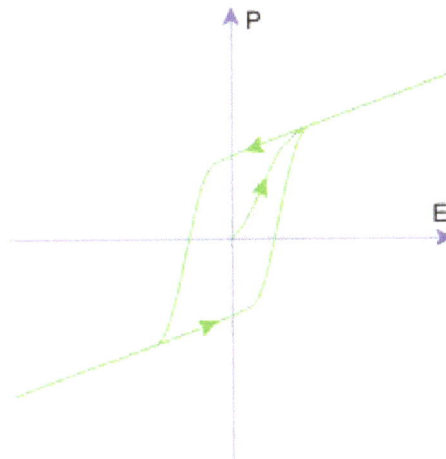

A typical ferroelectric hysteresis loop

Principle of Ferroelectricity: Energetics

The movement of central atom in the above structure can explained in terms of a potential energy diagram.

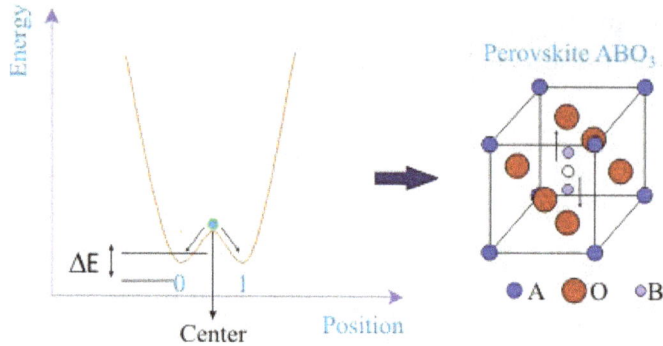

Potential energy well and crystal structure for an ABO3 structured ferroelectric

This situation can be explained in terms of a potential energy between two adjacent low energy sites. There are two equilibrium positions in which a B ion can stay, but to change from one state to another, energy must be provided to overcome an energy barrier ΔE. These energy wells further tilt to the left or right, depending upon the polarity of the electric field i.e. in the non-zero field state making on configuration more stable than another. However, a coercive field, which is the field required to bring the polarization back to zero, is needed when you switch to other direction of the field.

Ferroelectric materials follow Curie-Weiss law which is expressed as

$$x = \frac{\dfrac{N\alpha}{\varepsilon_0}}{1 - \dfrac{N\alpha}{3\varepsilon_0}} = \frac{3T_c}{T - T_c}$$

Here, N is the number of dipoles per unit volume of the material, α is dipolar polarizability, and T_c is defined as the Ferroelectric transition temperature or the Curie temperature.

Ferroelectric behaviour is observed below the Curie Temperature above which the ferroelectric phase converts into a paraelectric phase which always has higher symmetry than the ferroelectric phase (for example transformation of low symmetry tetragonal $BaTiO_3$ to higher symmetry cubic $BaTiO_3$ at about 120°C while heating) and ferroelectricity disappears. A paraelectric state is essentially a centrosymmetric higher symmetry state where dipoles are randomly oriented in a crystal giving rise to zero polarization.

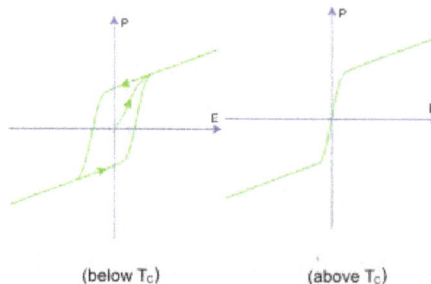

(below T_c)   (above T_c)
Schematic of Ferroelectric to Paraelectric transition

Proof of Curie-Weiss Law

We can now say that in a ferroelectric material, reversible spontaneous alignment of electric dipoles takes place by mutual interaction. This happens due to a local field E' which is increased by dipole alignment in the direction of field. Interestingly, this phenomenon happens below a critical temperature T_c when crystal enters into nonsymetric state where thermal energy cannot randomize the electric dipoles, presumably due to dipole-dipole interactions and local field.

Polarization P can be expressed as:

$$P = (\varepsilon_r' - 1)\varepsilon_0 E = N\alpha E'$$

Where local field E' = E + P/3ε_0 as given by Clausius-Mossotti relationship and e_r' is the relative dielectric permittivity and α is total polarizability. Substitution for E' in above equation results in

$$P = N\alpha \left(E + \frac{P}{3\varepsilon_0} \right) = N\alpha E + \frac{N\alpha P}{3\varepsilon_0}$$

i.e.

$$P \left(1 - \frac{N\alpha}{3\varepsilon_0} \right) = N\alpha E$$

or

$$P \frac{N\alpha E}{\left(1 - \dfrac{N\alpha P}{3\varepsilon_0} \right)}$$

Since we know that susceptibility, $\chi = \varepsilon_r' - 1 = \dfrac{P}{\varepsilon_0 E'}$, we get

$$\chi = \varepsilon_r' - 1 = \frac{(N\alpha/\varepsilon_\circ)}{\left(1 - \dfrac{N\alpha}{3\varepsilon_\circ} \right)}$$

Equation above shows that both χ and ε_r' must approach ∞ when Nα/3ε_0 approaches 1.

This can be assumed to be a right condition for ferroelectrics as near the ferroelectric transition, they exhibit very large susceptibilities and dielectric constant. At this point, we can also ignore the electronic, ionic and interface polarizations assuming that dipolar polarizability is too high such that $\alpha_d \gg \alpha_e + \alpha_i + \alpha_{int}$ at a critical temperature T_c. Here define $\alpha = \alpha_d = C/kT$ where C is Curie's constant. This is the right sort of relation as we have already seen the temperature dependence of dipolar polarization.

Hence, we can further write

$$\frac{N_\alpha}{3\varepsilon_\circ} = \frac{N}{3\varepsilon_\circ}\left(\frac{C}{kT_c}\right) = 1$$

OR

$$T_c = \frac{NC}{3k\varepsilon_\circ}$$

Below this T_c spontaneous polarization is prevalent and dipoles tend to align. So, since $C = \alpha_d kT$, we get

$$T_c = \frac{N\alpha_d T}{3\varepsilon_\circ} = \frac{N\alpha T}{3\varepsilon_\circ}$$

OR $\qquad \dfrac{T_c}{T} = \dfrac{N\alpha}{3\varepsilon_\circ}$

Thus, by now modifying equation, Curie-Weiss law can be expressed as

$$\chi = \frac{N\alpha/\varepsilon_\circ}{1 - \dfrac{N\alpha}{3\varepsilon_\circ}} = \frac{3T_c/T}{1 - \dfrac{T_c}{T}} = \frac{3T_c}{T - T_c}$$

Thermodyanamic Basis of Ferroelectric Phase Transitions

Thermodynamic theory to understand the ferroelectric phase transitions was developed after contributions from Lev Landau and V.L. Ginzburg, both Soviet physicists and A.F. Devonshire, British Physicist. The approach is based around calculating the free energy of system and working around other thermodynamic parameters to predict the nature of phase transition. Here, before we go into details of this theory, we will look at the some of the basic definitions.

These non-linear dielectrics exhibit various kinds of couplings between physical properties and can be expressed mathematically. For instance, below are the expressions for ferroelectric, piezoelectric and pyroelectric couplings.

Ferroelectric Effect:

Electric charge in a polar material can be induced by application of an external electric field and can be expressed as

$$\overline{P}_1 = \chi_{ij} E_j$$

where χ_{ij} is the susceptibility in F/m (actually dimensionless but here $\varepsilon_\circ$ i.e. permittivity of free space is also included which has dimension of F/m) and is second rank tensor. *Note that the equation is valid only for the linear region of the hysteresis curve.*

Piezoelectric effect:

Similarly, the charge induced by application of external stress i.e. piezoelectric effect, can be expressed by

$$D_i = d_{ijk}\sigma_{jk}$$

where d_{ijk} is the piezoelectric coefficient and is third rank tensor with units C/N, σ_{jk} is the stress applied.

Converse piezoelectric effect is expressed as

$$x_{ij} = d_{kij}E_k = d^t_{ijk}E_k$$

Where d_{ijkt} is in m/V.

Pyroelectric Effect:

Induced charge by temperature change i.e. pyroelectric effect is expressed by

$$P_i = \frac{\partial PS_i}{\partial T}$$

where P_i is the vector of pyroelectric coefficient in $cm^{-2}K^{-1}$.

Displacement is expressed as

$$D_i = \Delta P_{Si} = p_i\Delta T$$

So, depending upon the state of material, many such effects may be present together. These couplings between thermal, elastic or electric properties can be understood formally by adopting a thermodynamic approach. The results of such an approach yield equations of state which relate the material parameters with different experimental conditions which assist in modeling of the parameters and in understanding the response of various devices.

From the laws of thermodynamics, the thermodynamic state of any crystal in a state of equilibrium can be completely established by the value of number of variables, which in case of ferroelectrics include temperature T, entropy S, electric field E, polarization P, stress s and strain e. Usually parameters like electric field E and stress s can be treated as external or independent variables while polarization and strain can be treated as internal or dependent variables.

For a ferroelectric system, the free energy G can be expressed in terms of ten variables as

$$G = f(P_x, P_y, P_z, \sigma_x, \sigma_y, \sigma_z, \sigma_{yz}, \sigma_{zx}, \sigma_{xy}, T)$$

where P_x, P_y, P_z are the components of the polarization, $\sigma_x, \sigma_y, \sigma_z, \sigma_{yz}, \sigma_{zx}, \sigma_{xy}$ are the stress components and T is the temperature.

We can get the value of the independent variables in thermal equilibrium at the free energy minimum. For an uniaxial ferroelectric, free energy can be expanded in terms of polarization ignoring the stress field. Here, we select the origin of free energy for a free unpolarized and unstrained crystal to be zero. Hence,

$$G = \frac{1}{2}a\,P^2 + \frac{1}{4}b\,P^4 + \frac{1}{6}cP^6 + ... - EP$$

Note that only even powers are taken because energy is same for $\pm P_s$ states.

Here a, b, c are the temperature dependent constants and E is the electric field. The equilibrium is found by establishing $\left(\dfrac{\partial G}{\partial p}\right) = 0$ i.e.

$$aP + bP^3 + cP^5 + ... - E = 0$$

i.e.

$$E = aP + bP^3 + cP^5$$

If all of a, b, and c are positive, then P = o is the only root of the equation as shown below in the figure. This is situation for a paraelectric material where polarization is zero when field is zero.

If we ignore higher power terms, then

$$\frac{\partial F}{\partial P} = 0 = aP - E$$

leading to

$$\chi = a^{-1} = \frac{P}{E}$$

which is the same expression that we encountered for linear dielectrics.

According to Landau-Devonshire theory, near the Curie point (T~T_o) we assume

$$a = a_o(T - T_o)$$

As a result, the free energy expansion, only 'a' is dependent on temperature while other constants are temperature independent. Incorporating above equations

$$G = \frac{1}{2}a_o(T - T_o)P^2 + \frac{1}{4}bP^4 + \frac{1}{6}cP^6 + ... - EP$$

In this expression, for all known ferroelectrics, both ao and c are positive while depending upon the sign of b, the phase transition nature changes.

Figure below (a) show the free energy vs polarization plot when $T >> T_0$ i.e. when the crystal is in paraelectric state. On the other hand if, $a < 0$ when $T << T_0$ and b and c are positive, there will be a nonzero root of P in addition $P = 0$, as shown below in figure (b), representing the ferroelectric state with non-zero polarization at zero field.

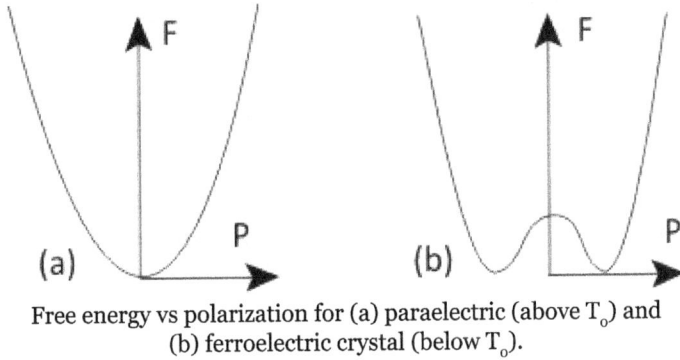

Free energy vs polarization for (a) paraelectric (above T_0) and (b) ferroelectric crystal (below T_0).

Case I: Second Order Transition

When b is positive, the ferroelectric transition occurs at a temperature $T = T_0$ and is called as second order transition (do not get confused between T_0 and T_c as the distiction will become clear in the first order phase transition). Under such a situation, free energy as function of polarization evolves continuously when temperature is changes i.e. from a curve with single minima at $P = 0$ when $T < T_0$ to a plot with two minima at $P = +P_0$ and $-P_0$ when $T > T_0$. Two curves become closely related if 'a' changes continuously with temperature and changes sign at T_c and can be shown together as in figure (a) below.

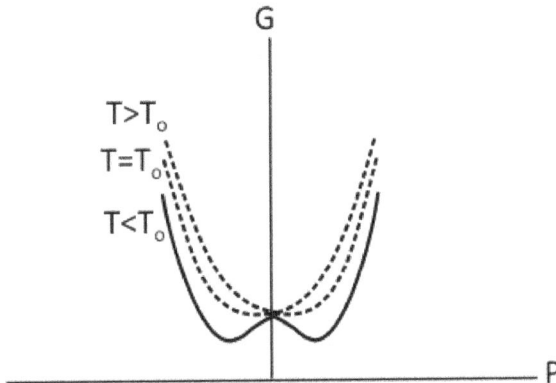

Effect of temperature on the free energy vs polarization plot. Note how the sign of a changes with temperature and its effect on the curve.

The spontaneous polarization, P_0, can be estimated by substituting $E = 0$ in equation (20) and retaining only two lowest order terms since all the coefficients i.e. a_0, b and c are positive. The polarization can be expressed as

$$P_0 = \frac{a_0}{b}(T_0 - T)^{1/2}$$

showing that the polarization decreased to zero at T = T$_0$ as shown in figure.

Dielectric susceptibility at T < T$_0$ can be estimated as

$$\chi = \frac{1}{2a_\circ}(T_\circ - T)^{-1}$$

Showing that susceptibility will have a divergence at T = T$_0$ or its reciprocal (i.e. dielectric stiffness) will vanish at T = T$_0$ as shown in figure 5.8. In real materials, susceptibility reaches very large values near T$_0$.

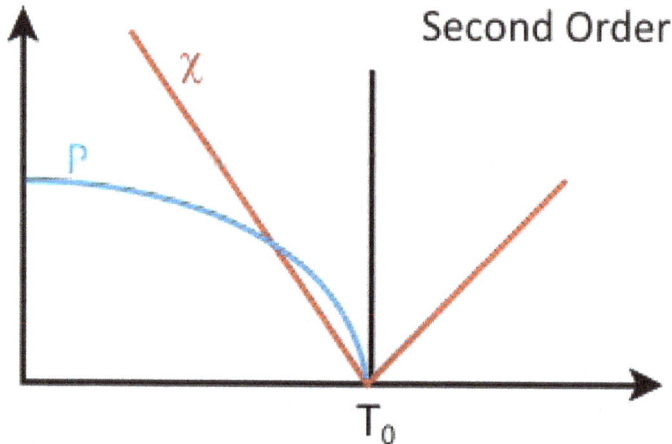

Effect of temperature on polarization and reciprocal susceptibility
for a second order phase transformation.

This transition is also depicted by a discontinuity in specific heat at transition which is estimated by using P = 0 at T > T$_0$ while using value given as T < T$_0$.

$$\Delta C_v = C_v(T = T_\circ^+) - C_v(T - T_\circ^-)$$

Now substituting $C_v = -T\dfrac{\partial^2 G}{\partial T^2}$ yields

$$\Delta C_v = \frac{a_\circ^2 T_\circ}{2b}$$

Examples of ferroelectric materials showing a second order transition are materials like Rochelle salt and KH_2PO_4.

Case – II: First Order Transition

Another situation to consider is that when a < 0, b < 0 but c > 0. What this means is that free energy vs polarization plot has three equal minima, one for P = 0 and the other two for P ≠ 0 at the same temperature i.e. at the same value of 'a' at a temperature T = T$_0$, Curie temperature, which is now more than T$_0$. This gives rise to the following free energy vs polarization plot.

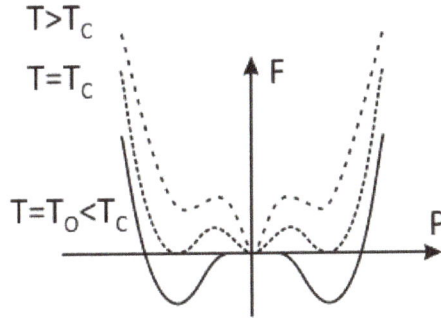

Free energy vs polarization schematic plot for a first order phase transition

The most important feature of this phase transition is that polarization i.e. the order parameter drops from $P \neq o$ to zero discontinuously at $T = T_c$ and is called as first order phase transition. This is also very clearly demonstrated by a discontinuity in the reciprocal of dielectric susceptibility as seen below. For example, solid-liquid phase transition is a first order phase transition while among various ferroelectrics, barium titanate is a fine example of first order transition among ferroelectrics.

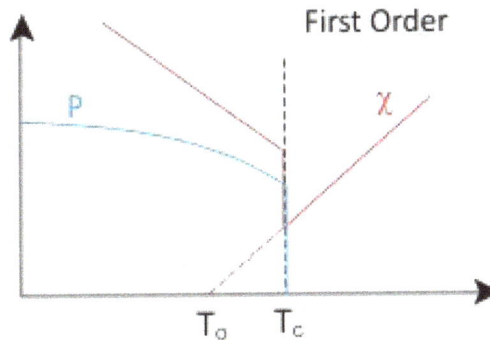

Polarization and reciprocal susceptibility plot for a first order phase transition

In order to compute the non-zero polarization (P_o) and susceptibility at the transition, the value of free energy for $P = o$ and $P = P_o$ must be equal at $T = T_c$ i.e.

$$G(0) = \frac{1}{2}aP^2 + \frac{1}{4}bP^4 + \frac{1}{6}cP^6 = 0$$

On the other hand, field E must also be zero for the polarization to be spontaneous i.e.

$$\frac{dG}{dP} = E = aP + bP^3 + cP^5$$

The polarization and susceptibility at T_c are obtained by solving two equations and are given as

$$P_o^2(T_c) = \frac{3}{4}\left(-\frac{b}{c}\right)$$

and

$$\left(\chi(T_c)\right)^{-1} = a = \frac{3}{16}\left(\frac{b}{c}\right)^2$$

The fact that there are three minima at $T=T_c$ is reflected in whether the T_c is approached while heating or cooling. More specifically, the material will be in other of the two non-zero polarization states if is heated from an initial temperature that is lower than T_c whereas, if it is cooled from a temperature higher than T_c, the sample will be in paraelectric state. This results in thermal hysteresis when these materials are thermally cycled across T_c.

Ferroelectric Domains

In a ferroelectric crystal, it is likely that the alignment of dipoles in one of the polar directions extends only over a region of the crystal and there can be different regions in the crystal with aligned dipoles which are oriented in many different directions with respect to one another.

Regions of uniform polarization are called domains, separated by a boundary called domain wall. You should not confuse ferroelectric domain walls with the grain boundaries. Depending upon the grain size, one grain can have more than one or more domains.

The types of domain walls that can occur in a ferroelectric crystal depend upon the crystal structure and symmetry of both paraelectric and ferroelectric phases. For instance, rhombohedral phase of lead zirconate titanate, $Pb(Zr,Ti)O_3$ has P_s vector along [111]-direction which gives eight possible directions of spontaneous polarization with $180°$, $71°$ and $109°$ domain walls. On the other hand, a tetragonal perovskite like $PbTiO_3$ has P_s along the [001]-axis and here domain walls are either $180°$ or $90°$ domain walls.

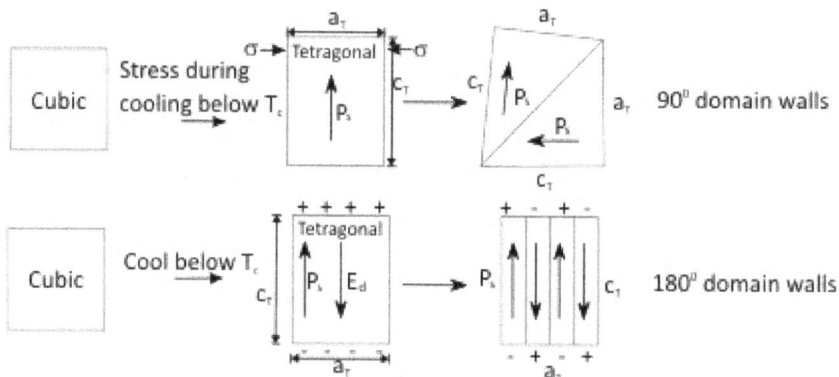

Schematic representation of a $180°$ and $90°$ domain walls in a tetragonal perovskite crystal such as $BaTiO_3$

Formation of the domains may also be the result of mechanical constraints associated with the stresses created by the ferroelectric phase transition e.g. from cubic paraelectric phase to tetragonal paraelectric phase in $PbTiO_3$. Both $180°$ and $90°$ domains minimize the energy associated with the depolarizing field but elastic energy is minimized only by the formation of $90°$ domains. Combination of both effects leads to a complex domain structure in the material with both $90°$ and $180°$ domain walls.

Domain wall

The driving force for the formation of domain walls is the minimization of the electrostatic energy of the depolarizing field (E_d), due to surface charges due to polarization, and the elastic energy associated with the mechanical constraints arising due to ferroelectric-paraelectric phase transition. This electrostatic energy associated with the depolarizing field can be minimized by

- splitting of the material into oppositely oriented domains or

- compensation of the electrical charge via electrical conduction through the crystal.

Domains can also be seen by microscopy. The following is an image of domains in BaTiO$_3$ as seen by transmission electron microscopy.

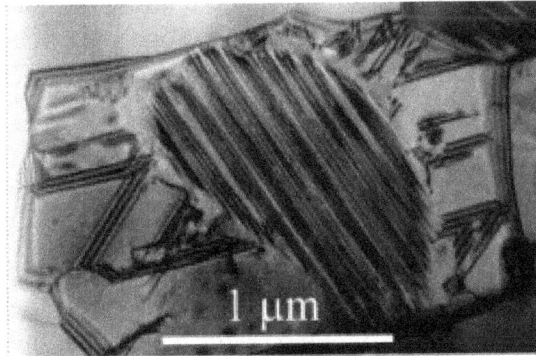

Domains in BaTiO$_3$ samples as seen by TEM

Analytical Treatment of Domain Wall Energy

Ferroelectric domains form as a result of instabilities due to alignment of dipoles in a crystal. The free energy change involved in the formation of a domain is given as

$$\Delta G = (G - G^\circ) = U_c + U_P + U_x + U_w + U_d$$

where

U_c : effect of applied field on the domain energy

U_p and U_x : bulk electrical and elastic energies

U_d : depolarization energy and

U_w : domain energy

U_d is the energy related to the internal field set up in the crystal by the polarization and not compensated. The internal field opposes the applied field E and hence is called as depolarizing fieldThis is dependent on the domain size (d) and is expressed as

$$U_d = \frac{\epsilon^* . d . V . P_0^2}{t}$$

Here ϵ^* is a constant determined by the dielectric constant of the material, t is crystal thickness, P_0 is the polarization at the center of the domain.

U_w is the domain energy and can be expressed in terms of surface energy of the wall (γ), domain width (d) and crystal volume (V) and is given as

$$U_w = \gamma . \frac{d}{V}$$

To get a stable domain structure, this ΔG has to be minimized. U_p and U_x are the same in each domain, irrespective of domain thickness and so are independent of domain size and hence are treated as constant. Therefore one only needs to consider U_w and U_d when ΔG is minimized w.r.t. domain wall thickness d i.e. $d\Delta G/dd = 0$ resulting in

$$d = \left(\frac{\gamma t}{\epsilon^* . P_0^2} \right)^{1/2}$$

Here, one can see that domain size is dependent on surface energy, crystal thickness and polarization, showing a competition between the surface energy of the wall and polarization (governing the depolarizing field). This shows that larger the surface energy is, larger the domain size is which makes sense because of large energy requirement, the interface area needs to be smaller. Secondly, the larger the polarization is, the larger the depolarizing field will be and large would be the driving force for the domains to form and hence smaller domains will be preferred.

Ferroelectric Switching and Domains

Application of an electric field to a ferroelectric ceramic leads to the alignment of anti-parallel or off-aligned dipoles to reorient themselves along the field.

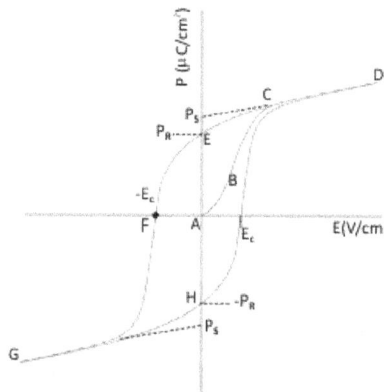

Characteristic hysteresis loop of a ferroelectric material

At sufficiently large fields, all the dipoles will be aligned along the field and the direction reverses by reversing the direction of the field. This phenomenon of polarization reversal takes place by way of nucleation and growth of favourably oriented domains into the unfavourably oriented domains and associated domain wall motion.

If we assume that our hypothetical crystal has an equal number of positive and negative domains in the virgin states, then the net polarization of the crystal will be zero. Now what happens when field E is applied? The plot that we get is something like shown above where P is polarization in $\mu C/cm^2$ and E is electric field across the sample in V/cm. The process is something like this:

Initial polarization P increases linearly with the increasing electric field and the crystal behaves like a dielectric because the applied field is not large enough to switch any of the domains oriented opposite to its direction. This linear region is shown as AB.

Further increase in the field strength forces nucleation and growth of favourably oriented domains at the expense of oppositely oriented domains and polarization starts increasing rapidly (BC) until all the domains are aligned in the direction of the electric field i.e. reach a single domain state (CD) when polarization saturates to a value called saturation polarization (P_S). The domains reversal actually takes place by formation of new favourably oriented domains at the expense of unfavourable domains. Now, when the field is decreased, the polarization generally does not return to zero but follows path DE and at zero field some of the domains still remain aligned in the positive direction and the crystal exhibits a remanent polarization (P_R). To bring the crystal back to zero polarization state, a negative electric field is required (along the path EF) which is also called the coercive field (E_C).

Further increase of electric field in the opposite direction will cause complete reversal of orientation of all domains in the direction of field (path FG) and the loop can be completed by following the path GHD. This relation between P and E is called a ferroelectric hysteresis loop which is an important characteristic of a ferroelectric crystal. The principle feature of a ferroelectric crystal is not only the presence of spontaneous polarization but also the fact that this polarization can be reversed by application of an electric field.

Measurement of Hysteresis Loop

Ferroelectric hysteresis loops can be experimentally measured using a Sawyer-Tower circuit, using a high frequency ac field, and are observed on the screen of an oscilloscope.

The circuit is schematically drawn below. A linear capacitor C_0 is connected in series with the crystal. In this configuration, the voltage across C_0 is proportional to the polarization of the crystal. This circuit not only measures the hysteresis loop, it also quantifies the spontaneous polarization P_s and coercive field E_c.

A ferroelectric materials shows polarization of the order of 50-100 $\mu C/cm^2$.

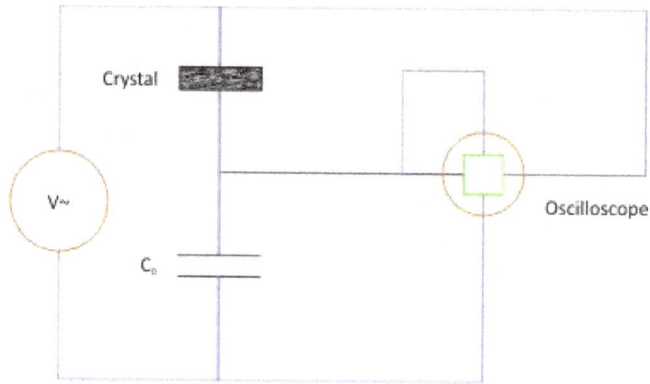

Schematic representation of sawyer-tower circuit

Nowadays, one can get commercially available instruments such as Radiant Premier Precision Station from Radiant Technologies which can perform polarization measurements at varying field and frequencies. For device measurements, one needs to establish top and bottom contacts to the ferroelectric materials which is typically achieved using thin platinum electrodes in thin films and silver paste in bulk ceramics.

Structural Change and Ferroelectricity in Barium Titanate (BaTiO$_3$)

Barium Titanate (BaTiO$_3$) has a perovskite ABO$_3$ type structure. As shown below, the central Ti atom is surrounded by six oxygen ions in a octahedral co-ordination determined by the radius ratio.

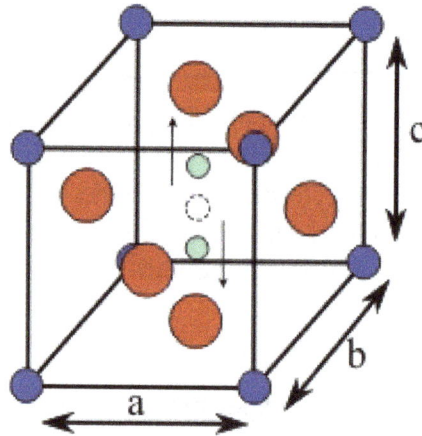

●A ●O ○B

Structure of Barium Titanate (A: Ba, B: Ti). For Cubic form
a=b=c while for tetragonal a=b≠c.

Above 120°C, cubic form of (BaTiO$_3$) has regular octahedrons of O^{2-} ions around Ti^{4+} ion and has a center of symmetry. As a result, the six Ti-O dipole moments along $\pm x$, $\pm y$, $\pm z$ cancel each other and the material in such a state is called paraelectric.

Below 120°C, BaTiO$_3$ transforms to a noncentrosymmetric tetragonal phase with one of the axis becoming longer, typically referred as z-axis or [001]-direction. Unilateral displacement of the positively changed Ti^{4+} ions against surrounding O^{2-} ions occurs to give rise to net permanent dipole moment. Coupling of such displacements and the associated dipole moment is a necessity for ferroelectricity. This transformation forces Ti ions go to lower energy off center positions, giving rise to permanent dipoles.

The crystallographic dimension of the BaTiO$_3$ lattice change with temperature as shown below.

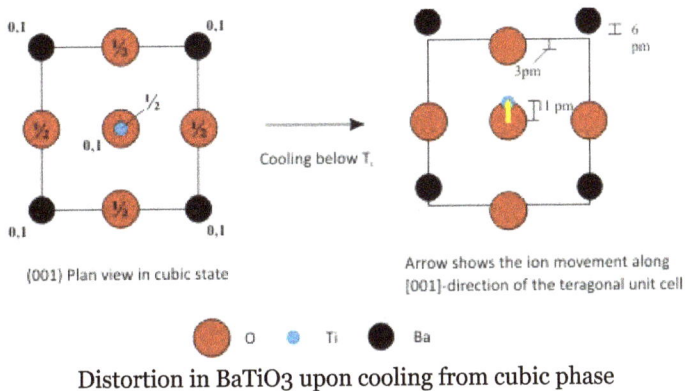

Cooling below T$_c$

(001) Plan view in cubic state

Arrow shows the ion movement along
[001]-direction of the teragonal unit cell

O   Ti   Ba

Distortion in BaTiO3 upon cooling from cubic phase

Because the distorted octahedra are coupled together in ferroelectric form, there is a very large spontaneous polarization, ~25 µC/cm², giving rise to a large dielectric constant, ~160, and large temperature dependence of dielectric constant.

BaTiO$_3$ shows two more structural transitions when cooled further below 120°C. It transforms to orthorhombic structure at ~5°C and then again to a rhombohedral structure at ~ -90°C and as result of change in the symmetry, the polarization vector also changes from [001] for tetragonal to [110] in orthorhombic and [111] in rhombohedral structure.

Applications of Ferroelectrics

In addition to BaTiO$_3$, extensively studied ferroelectric materials have been PbTiO$_3$, Pb(Zr,Ti)O$_3$, Bi$_4$Ti$_3$O$_{12}$ and SrBi$_2$Ta$_2$O$_9$. While the first two show large polarization and a reasonably high T$_c$ (above 400°C), the latter two do not contain lead and have higher Curie transition temperatures.

Although many applications of ferroelectrics are for their piezoelectric properties, ferroelectrics can be used for applications like non-volatile data storage below their T$_c$. Above T$_c$ where their dielectric constant increases linearly with temperature, they can be used for camera flashes.

1. Nonvolatile Memories

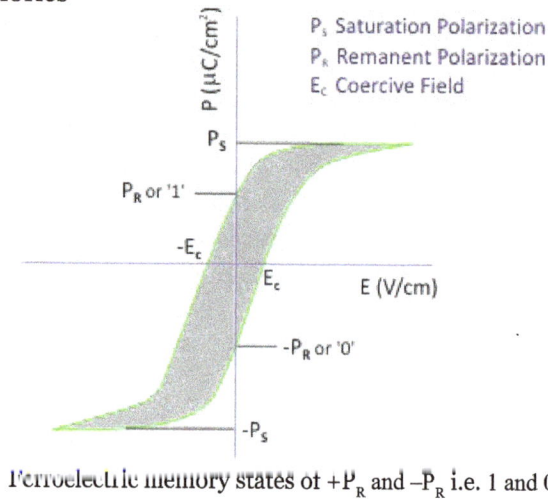

Ferroelectric memory states of $+P_R$ and $-P_R$ i.e. 1 and 0

Since ferroelectric materials show a hysteresis loop and remnant polarization $\pm P_R$ at zero field, these two polarization states can be used as '0' and '1' states of binary data storage in memory devices. The advantages are that data will be stored when power is lost during operation leading to non-volatile data storage. Other advantages are that ferroelectric switching is a very fast phenomenon and hence memories can operate very fast. Many of these materials tend to be radiation resistant and hence can be used in space applications.

2. Camera Flashes

In this application, the battery charges the ferroelectric capacitor first. Then, once fully charged, the ferroelectric is connected to the bulb and causes it to flash.

Piezoelectric Ceramics

Piezoelectric effect was discovered by Jacques and Pierre Curie in 1888. Direct piezo-electric effect is the ability of some materials to create an electric potential in response to applied mechanical stress. The applied stress changes the polarization density within the material's volume leading to the observed potential. As a requirement, only materials with non-centrosymmetric crystal structure can exhibit piezoelectric effect. Some of the commonly used/known piezoelectric materials are quartz (SiO_2), zinc oxide (ZnO), polyvinylidenefluoride (PVDF) and lead zirconate titanate, (PZT or $Pb(Zr,Ti)O_3$).

An oscillating applied stress on a piezoelectric material can give rise to the field which can be applied to an electrical load such as a bulb. Another example can be charging of your mobile or any other device in your backpack while you walk. You could not achieve the same while standing.

For a detailed discussion on the piezoelectric properties, materials, and applications, readers can refer to the bibliography provided in the beginning.

Direct Piezoelectric Effect

Direct effect occurs when an applied stress to a material gives rise to a change in the polarization density which in turn can be detected as electric field or potential across the sample. Here, the polarization is directly proportional to the stress applied, as described by the equation.

$$P = d.\sigma$$

where P is polarization, σ is applied stress and d is piezoelectric coefficient (actually a third rank tensor).

Reverse or Converse Piezoelectric Effect

The reverse is true is when an electric field is applied to the material and as a result, a strain is induced expressed as

$$\varepsilon = d.E$$

where ε is the strain induced, d is the piezoelectric coefficient and E is the applied electric field.

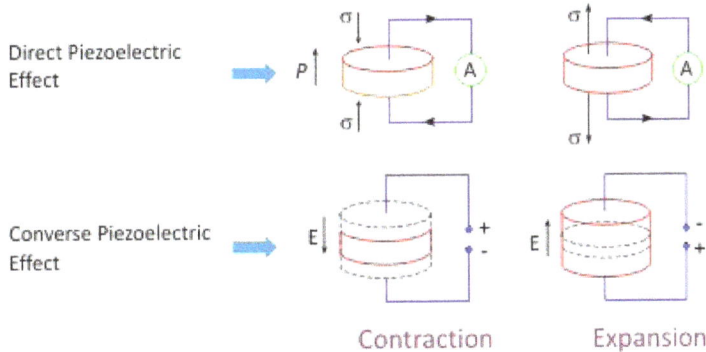

Direct and Converse Piezoelectric Effects

The direct piezoelectric effect is used as the basis for force, pressure, vibration and acceleration sensors while converse effect is used as a basis for actuator and displacement devices.

Poling of Piezoelectric Materials

There are some piezoelectrics such as quartz which are not spontaneously polarized but get polarized upon application of stress, while ferroelectric which are anyway piezoelectric in nature are spontaneously polarized and show a change of polarization upon application of stress.

The values of piezoelectric coefficient of some materials are given below:

Material	Piezoelectric Constant, d (pm/V)
Quartz	2.3
Barium Titanate	100-149
Lead Niobate	80-85
Lead zirconate titanate	250-365

So, you can observe from this table that the level of strain generated is not so massive but is still important because of preciseness and reversibility of the effect.

Most ferroelectrics have to be poled to be useful as a piezoelectric. In the unpoled virgin state of the material, the ferroelectric domains of single polarization direction are randomly distributed across the material and in such a situation the net polarization would be zero. Application of stress to such a material would not achieve any change in the net polarization, thus making it useless as a piezoelectric.

Poling i.e. application of a large electric field near T_c (just below T_c) orients the domains along the field and when the field is removed, the domain structure does not get back to the original condition giving rise to a net polarization along a certain direction. Now, when stress is applied to such a crystal, a noticeable change in the polarization can be observed.

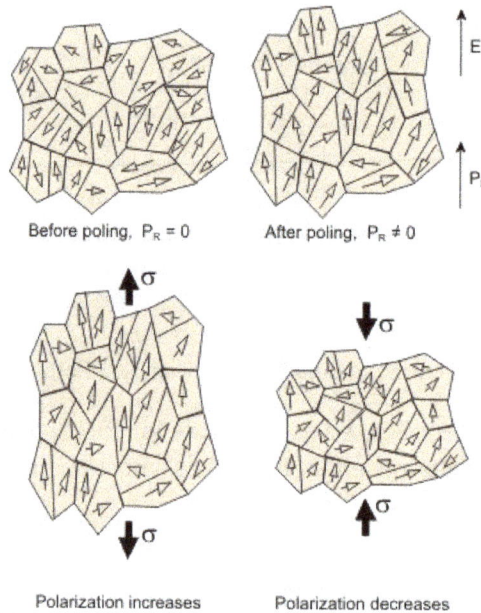

Before poling, $P_R = 0$ After poling, $P_R \neq 0$

Polarization increases   Polarization decreases

Poling of ferroelectrics and application of stress on poled material

Depolarization of Piezoelectrics

Just like a ferroelectric material can be poled, opposing polarity, high electric field or thermal cycling close to T_c or the application of large mechanical stresses can lead to the disappearance of polarization or rather result in alignment of dipoles gets lost.

Common PIezoelectric Materials

1. Barium Titanate ($BaTiO_3$)

This was the first piezoelectric material which was developed commercially for application in the generation and detection of acoustic and ultrasonic energy. The transition temperature can be modified by chemical substitutions:

Ba substitution by Pb and Ca lowers the T_c of tetragonal to orthorhombic transition. This has been used to control the piezoelectric properties around 0°C, and is important for underwater detection and echo sounding.

Ti substitution by Zr or Sn increases the transition temperature for both the tetragonal–orthorhomic and orthorhomic–rhombohedral transitions and enhances piezoelectric properties. Ti substitution by 1-2 at % Co^{3+} leads to much reduced losses as high fields useful in ultrasonic applications. Care must be taken during processing to avoid reduction of Co^{3+} to Co^{2+} which occurs very easily.

2. $Pb(Zr,Ti)O_3$ or PZT

PZT is one of the most used piezoelectric in a variety of applications due to its excellent properties and high enough transition temperatures. It has a perovskite structure with B sites randomly occupied by either of isovalent Ti and Zr ions.

The phase diagram of PZT is shown below.

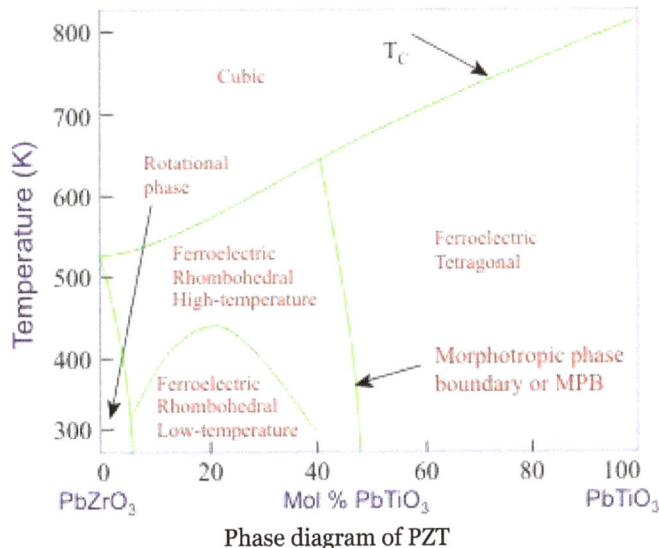

Phase diagram of PZT

The typically used composition is about Zr:Ti::50:50 which gives excellent properties. The reason for this is that this is closed to morphotropic phase boundary and here rhombohedral and tetragonal structures co-exist. As we know that P_s vector is along [001]-direction in tetragonal phase and <111>- direction in rhombohedral phase, it permit material to be poled easily as there are quite a few poling directions available making it a useful piezoelectric.

Donor doping in PZT such as La^{3+} on Pb site reduces the concentration of oxygen vacancies. This in turn reduces the concentration of defect pairs which otherwise impede the domain wall motion. This leads to noticeable increases in the permittivity, dielectric losses, elastic compliance and coupling coefficients, and reduction in the coercivity.

Measurement of Piezoelectric Properties

Piezoelectric measurements are usually made to measure the displacement of the material when an electric field is applied. These techniques are resonance or subresonance techniques.

In the resonance methods, one conducts the measurement of the characteristic frequencies of the materials upon the application of alternating electric field and is widely used for bulk samples. To a first approximation, the electromechanical response of a piezoelectric material close to the characteristics frequency can represented by the electrical equivalent circuit as shown in the figure. Simplest measurements are conducted by poling a piezoelectric long rod of length ~6 inch and diameter ~¼ inch along its length.

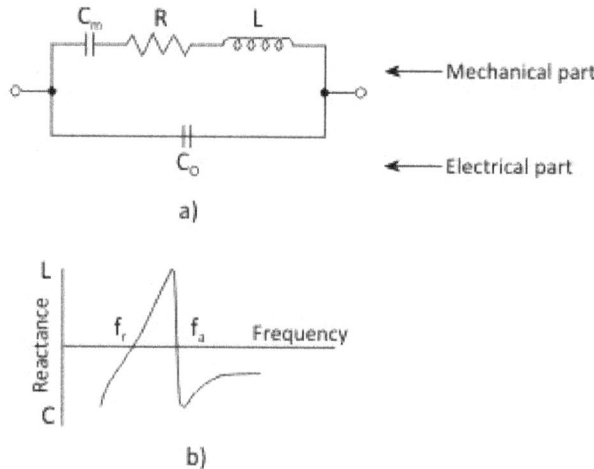

a)

b)

Schematic representation of (a) equivalent electrical circuit of a piezoelectric sample close to its characteristic frequency (b) Plot of electrical reactance of the sample a function of frequency

The coupling coefficient, k_{33}, is expressed in terms of series and parallel resonance frequencies (f_s and f_p respectively) as

$$k_{33}^{2} = \frac{\pi f_s}{2 f_p} \tan\left(\frac{\pi}{2} \frac{\left(f_p - f_s \right)}{f_p} \right)$$

Using this relation along with elastic compliance and low frequency dielectric constant, the piezoelectric coefficient, d_{33}, can be calculated by using the following relation:

$$d_{33} = k_{33}.(S_{33}^E \times \varepsilon_{33}^X)^{\frac{1}{2}}$$

Measurements are limited to the specific frequencies determined by the fundamental vibration modes of the sample. In the case of piezoelectric thin films, the thickness resonance occurs in the GHz range in which the measurements involve considerable difficulties. In such cases, the resonance in the substrate driven by a thin film could be used to determine the piezoelectric coefficient of a thin film. The disadvantage of this method is that measurements are limited by the number of characteristics frequencies determined by the electromechanical response of a material.

One can use subresonance techniques for the measurement of piezoelectric properties at the frequencies which are much below the characteristics fundamental resonance frequencies. This includes measurement of direct effect *i.e.* charge developed on a piezoelectric material under application of an external mechanical stress and measurement of converse effect *i.e.* measurement of electric field induced displacements. Although displacements can be rather small to measure accurately, technological advances have allowed accurate measurements using techniques like strain gauges or linear variable differential transformers (LVDTs) or optical interferometers or atomic force microscopes. Appropriate electrical circuits needs to drawn and modeled to clearly elucidate the material properties.

Applications of Piezoelectric Ceramics

Piezoelectric ceramics are used in a variety of applications utilizing either direct or converse piezoelectric effect. The following are some applications of the piezoelectric ceramics:

Power Generation

- Gas Lighter

  Piezoelectric material can ignite the gases by generating a spark via an electric current. This requires two piezoelectrics with opposite polarization states which are brought close to each other so those polarization vectors are in the opposite directions i.e. faces containing similar charges are together. The piezoelectric are placed in a circuit with a spark gap.

  Now, application of a mechanical stress or force will induce change in the polarization. The force brings together these two pieces which then gives rise to creation of charges. The charges flow from the end faces and the middle (pressed) faces through the circuit giving rise to a spark in the spark gap which can be used to ignite a gas.

One must apply the force quickly otherwise the voltage generated disappears because the charges leaks away through the piezoceramic, across its surfaces and via the apparatus.

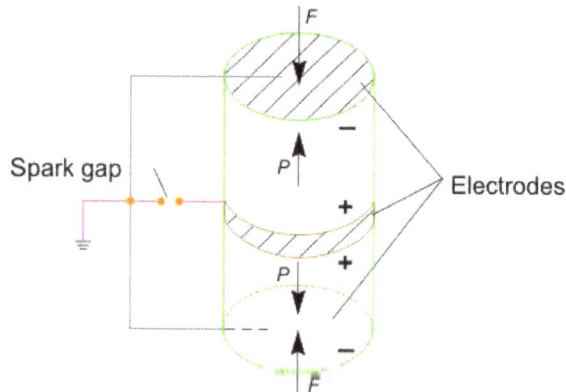

Schematic of operation of a gas lighter made using piezoelectric material

- Power Transformer

A piezoelectric transformer works like an AC voltage multiplier. While conventional transformers utilize magnetic coupling between input and output, the piezoelectric transformer exploits the acoustic coupling utilizing inverse piezoelectric effect. Piezo transformers can be quite compact high voltage sources.

An input smaller voltage across the thickness of a piezoceramic creates an alternating stress in the bar by the inverse piezoelectric effect. This causes the bar to vibrate with vibration frequency chosen to be the resonant frequency of the block, typically in the 100 kHz to 1 MHz range. This generates a higher output voltage in the other section of the bar by the direct piezoelectric effect. One can achieve the step-up ratios of more than 1000:1 using this technique.

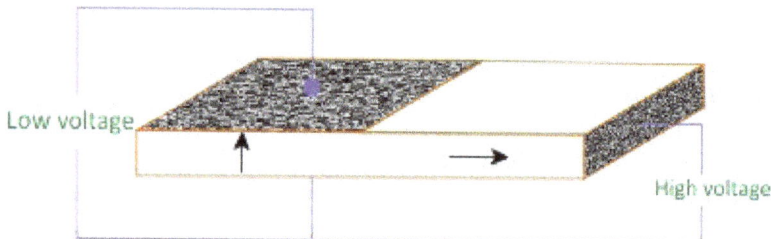

Schematic of a piezoelectric transformer

Piezoelectric Sensors

Here, typically pressure or force is used to create an electrical signal out of a piezoelectric material. For instance, in a microphone, sound waves can deform the piezoelectric element by bending it and thus giving a changing voltage. Similar principle can also be used for pickup guitars and microphones.

Other sensor applications are

- Detection and generation of sonar waves.

- To detect detonation in automotive engine by sampling the vibrations of the engine block

- To detect the precise moment of fuel injection in an automotive engine

- Detection of acoustic emissions in acoustic emission testing.

- Microbalances as very sensitive chemical and biological sensors.

- Strain gauges

- Medical applications using ultrasound waves

- Kidney stone treatment

In this application, electricity of high frequency is applied to the sample which gives rise to a change in the shape of the material. The shape change leads to emission of waves of frequencies in the ultrasound range. These powerful ultrasound waves can be used to shatter pieces of materials inside the body such as kidney stone, which can then pass out through the urine.

Ultrasound Imaging using Transduction Effect

Another application is its use in the ultrasound imaging of the fetus where piezoelectric acts as a transducer utilizing both direct and inverse effects. Since a high frequency field application to a piezoelectric can lead to emission of ultrasonic waves by direct effect, these waves, when they meet the tissues in the body, some of these waves get reflected. The reflected waves come back to piezoelectric exploit the inverse effect which leads to creation of charges from the piezoelectric which can then be modeled to generate an image of the fetus.

Actuators

In the precision engineering applications, precise linear or rotational movements are required for achieving technological perfection. In piezoelectrics, application of high electric fields (without using oscillations) correspond to only tiny changes in the crystal dimension and these changes can be very precise, achieving better than a micrometer precision. This ability makes these materials useful as precise actuators for achieving very precise motions.

In these applications, typically multilayer ceramics consisting of layers thinner than 100 microns, are used. One can achieve very high field in the multilayered materials using voltages lower than 150-200 V, not very high voltages.

You can have it in two forms:

- Direct Piezo Actuators with strokes lower than 100 microns or so and

- Amplified Piezoelectric Actuators which can yield millimeter long strokes.

Some of the examples of applications are

- Piezoelectric motors consisting of piezoelectric elements which apply a directional force to an axle, causing it to rotate. As the distances travelled are extremely small, it is a very high-precision replacement for the conventional stepper motor.

- Scanning force microscopes use inverse piezoelectric effect to keep the sensing needle close to the probe

- Laser mirror alignment in the laser electronics helping maintain accurate optical conditions inside the laser cavity to optimize the beam output.

- Loudspeakers: Voltage is converted to mechanical movement of a piezoelectric polymer film.

- In inkjet printers where piezoelectrics are used to control ink flow from the print head to the paper.

- As fuel injectors in diesel engines in place of commonly used solenoid valve devices.

Frequency Standards

Here quartz is most commonly used material as its piezoelectric properties are useful as standard of frequency.

- Quartz clocks

  Watches use a quartz tuning fork which uses a combination of both direct and converse piezoelectricity to give rise to a regularly timed series of electrical pulses that is used to mark time. The quartz crystal has a very accurately defined natural frequency of vibration at which it prefers to oscillate, and this is used to stabilize the frequency of a periodic voltage applied to the crystal. The same principle is critical in all radio transmitters and receivers, and in computers where it creates a clock pulse. Both of these usually use a frequency multiplier to reach the megahertz and gigahertz ranges.

Pyroelectric Ceramics

Pyroelectric materials possess a spontaneous polarization along a unique crystallographic direction which may or may not be reversible by changing the polarity of the applied field. If the latter is true, then a pyroelectric material is also ferroelectric. If it is ferroelectric material too, then the material can either be in a single crystalline state or in a poled state.

Pyroelectricity, in itself, is the ability of materials to generate a voltage when they are heated or cooled. It is temperature dependence of the spontaneous polarization in polar materials due to minute changes in the atomic positions as a result of change in the temperature. If the temperature is constant, then voltage gradually disappears due to leakage of charges through the material or air or the apparatus. Change in the polarization on a sample surface can be measured as an induced current.

Pyroelectricity was first observed by the Greek philosopher Theophrastus in 314 BC who found that Tourmaline attracted small pieces of straw and ash when it was heated. The first scientific description of this phenomenon was described by Louis Lemery in 1717. In 1747, Linnaeus first related the phenomenon to electricity, although this was not proven until Franz Ulrich Theodor Aepinus did so in 1756.

Difference between Pyroelectric and Ferroelectric Material

Although both ferroelectric and pyroelectric materials must be non-centrosymmetric and polar, the essential difference between them lies when an electric field is applied. While a change in temperature below Curie temperature leads to the creation of dipole along the polar axis by slight movement of atoms from their neutral positions (A), a reverse electric field can reverse the direction of polarization in a ferroelectric but not in a pyroelectric material (B). However, when the material is heated above Curie temperature, the atoms come back to their equilibrium positions (C).

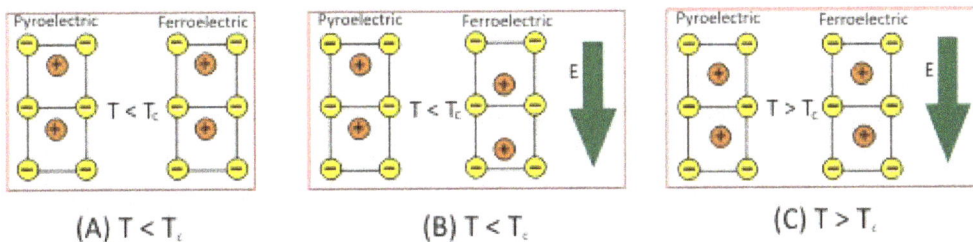

Difference between a ferroelectric and pyroelectric material

Theory of Piezoelectric Materials

From the fundamentals discussed, we can write that when an electric field E is applied to a material, the total dielectric displacement *i.e.* charge per unit area of the plates on both side of pyroelectric material *i.e.* D can be expressed as

$$D = \varepsilon E + P_s + P_{induced} = \varepsilon E + P_s$$

Here ε is permittivity of the pyroelectric material and P_s is the spontaneous polarization.

Assuming electric field, E, as constant, differentiating the above equation with temperature leads to

$$\frac{\partial D}{\partial T} = \frac{\partial P_s}{\partial T} + E\frac{\partial \varepsilon}{\partial T}$$

Now, defining generalized pyroelectric coefficient, Δp^g, as change in polarization with temperature, we write

$$\Delta p_g = \frac{\partial D}{\partial T} = \frac{\partial P_s}{\partial T} + E\frac{\partial \varepsilon}{\partial T} \qquad\qquad \text{Or}$$

$$\Delta p_g = p + E\frac{\partial \varepsilon}{\partial T}$$

Here p is defined as true pyroelectric coefficient. The last term in the above equation is the temperature dependence of the permittivity of the material which we can measure.

Since polarization is a vector, the pyroelectric coefficient is also a vector and has three components as defined by

$$\Delta P_i = p_i \Delta T$$

The above equation neglects the change in dielectric constant with temperature and is valid only when such assumption is true. However, in practice, the electrodes collecting the charges are normal to the polar axis, and we treat these quantities as scalars. The coefficient is usually negative because polarization decreases with temperature.

The behavior of polarization with temperature is dependent on the nature of transition, naturally the pyroelectric coefficients are much larger in magnitude for second order transition than for first order transition and hence are more useful.

Measurement of Pyroelectric Coefficient

One of the direct methods of measurement of pyroelectric measurement is shown below. The circuit connecting a pyroelectric material (held inside an oven) with an amplifier and then measuring the pyro-current. The pyroelectric coefficient is given as

$$p = \frac{I_p}{A.\frac{dT}{dt}}$$

Where I_p is the pyrocurrent and is given as

$$I_p = I_M\left(1 + \frac{R_s}{R_a}\right)$$

where R_s and R_a are the leakage resistance of the sample and input resistance of the amplifier respectively.

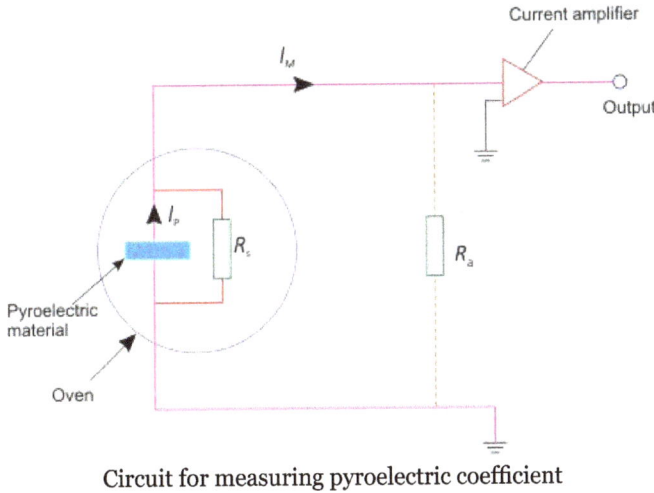

Circuit for measuring pyroelectric coefficient

Direct and Indirect Effect

Since all pyroelectric materials are naturally piezoelectric in nature, the thermal expansion/contraction of the material while heating or cooling induces thermal stresses in the material which in turn induce their own polarization and hence changing the overall polarization.

Since, change in the polarization can also be expressed as $\left(\dfrac{\partial P}{\partial T}\right) = \left(\dfrac{\partial P}{\partial T}\right)\left(\dfrac{\partial P}{\partial T}\right)\left(\dfrac{\partial P}{\partial T}\right)$,

depending upon the sign of each of these term, the polarization will either decrease or increase with increasing temperature.

Common Pyroelectric Materials

Most of the inorganic pyroelectrics (including ferroelectrics) are perovskite structured.

The most common materials are tabulated below.

Material	Structure	T_c (°C)	Pyroelectric Coefficient (μC.m^{-2}.K^{-1})
LiTaO$_3$ single crystal	Hexagonal	665	-230
0.75Pb(Mg$_{1/3}$-Nb$_{2/3}$)O$_3$-0.25 PbTiO$_3$ (PMN-PT) Ceramic	Perovskite	150	-1300
Ba$_{0.67}$Sr$_{0.33}$TiO$_3$ (BST) Ceramic	Perovskite	25	-7000
Triglycine sulphate (NH$_2$CH$_2$COOH)$_3$H$_2$SO$_4$	Sulphate	49	−280
Polyvinylidene fluoride (PVDF) film	Polymer	80	−27

Triglycine Sulphate (TGS)

- High pyroelectric coefficient

- Fragile and water-soluble, difficult to handle and cannot be used in devices where it would be subjected to either a hard vacuum or high humidity as it tends to decompose

- Can be modified to withstand temperatures above Curie point without depoling

- Used in thermal imaging cameras

Polyvinylidene Fluoride (PVDF)

- Poor pyroelectric coefficient

- Readily available in large areas of thin film

- More stable to heat, vacuum and moisture than TGS, mechanically robust

- Low heat conductivity and low permittivity

- High loss tangent

- Commonly used for burglar alarms

Perovskite Ferroelectric Ceramics

- Generally robust and insensitive to moisture and vacuum

- High pyroelectric coefficient and low loss

- Better operation near T_C

- Strong dependence on composition

- As a very approximate guide, for large area applications, low dielectric constant materials such as PVDF are preferred while for small area applications, materials with large dielectric constant such as perovskite oxides are preferred.

Common Applications

Burglar Alarms

Change in temperature of detector against the ambient temperature when an intruder comes in the vicinity of the alarm leads to a voltage which can be used to trigger an alarm. To avoid the effects due to thermal expansion, one needs to use a reference identical material to counter these extraneous effects. Such detectors can sense the objects' presence up to about 100 m.

Infrared or Thermal Imaging

Just like we use visible light to make a photograph, infrared (IR) radiation emitted by objects at different temperatures is focused onto a sensitive plate to create thermal image of the object. The atmospheric window typically used in IR imaging is from 8 to 14 µm and co-incidentally the power radiated from a black body at 300K peaks around 10 µm making it a perfect match.

The pyroelectric elements used in the devices are typically square plates with sides about a mm long and thicknesses around 30 µm. Because entire scenes are focused onto the plates in thermal imaging, they have to be larger, typically squares of side about 1 cm; the thicknesses are the same as for the simpler devices.

A typical photograph generated from IR imaging looks like this:

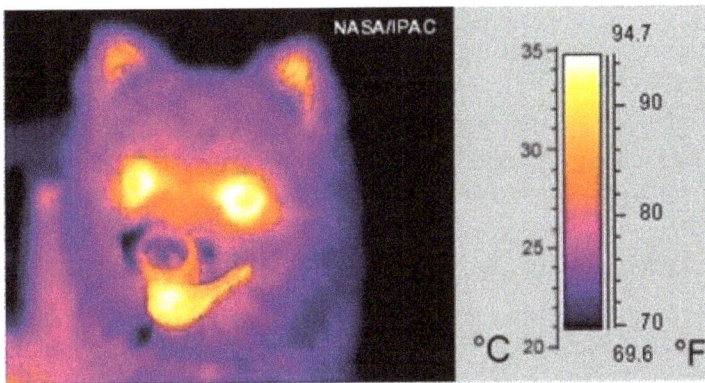

IR image of a dog. Bar on the right shows the temperature-colour relation

Here is a simple explanation of how IT imaging using pyroelectrics works.

A special lens focuses the infrared light emitted by all of the objects in view. The focused light is scanned by a phased array of pyroelectric elements which create a very detailed temperature pattern called a thermogram. It only takes about one-thirtieth of a second for the detector array to obtain to create a thermogram. This information is obtained from several thousand points in the field of view of the detector array.

This thermogram is translated into electrical signals which are sent to a signal-processing unit which then translates the information from the elements into data for the display. The signal-processing unit sends the information to the display, where it appears as various colors depending on the intensity of the infrared emission. The combination of all the signals from all of the elements creates the image.

Process of image creation

Pollutant Control

The reduction of pollution and greenhouse gases has become a major priority for our country as we progress. This requires us to monitor the levels of pollution. We can use pyroelectric materials for these purposes as well.

Pyroelectrics being excellent detectors of IR radiation, can easily detect the level of IR radiation which passes through a gas sample. Since each gas has a characteristic wavelength which it absorbs, we can measure it easily.

References

- Kuhn, U.; Lüty, F. (1965). "Paraelectric heating and cooling with OH—dipoles in alkali halides". Solid State Communications. 3 (2): 31. Bibcode:1965SSCom...3...31K. doi:10.1016/0038-1098(65)90060-8

- James, Frank A.J.L., editor. The Correspondence of Michael Faraday, Volume 3, 1841–1848, "Letter 1798, William Whewell to Faraday, p. 442.". The Institution of Electrical Engineers, London, United Kingdom, 1996. ISBN 0-86341-250-5

- Kong, L.B.; Li, S.; Zhang, T.S.; Zhai, J.W.; Boey, F.Y.C.; Ma, J. (2010-11-30). "Electrically tunable dielectric materials and strategies to improve their performances". Progress in Materials Science. 55 (8): 840–893. doi:10.1016/j.pmatsci.2010.04.004

- Kao, Kwan Chi (2004). Dielectric Phenomena in Solids. London: Elsevier Academic Press. pp. 92–93. ISBN 0-12-396561-6

- Lyon, David (2013). "Gap size dependence of the dielectric strength in nano vacuum gaps". IEEE Transactions on Dielectrics and Electrical Insulation. 20 (4). doi:10.1109/TDEI.2013.6571470

Permissions

All chapters in this book are published with permission under the Creative Commons Attribution Share Alike License or equivalent. Every chapter published in this book has been scrutinized by our experts. Their significance has been extensively debated. The topics covered herein carry significant information for a comprehensive understanding. They may even be implemented as practical applications or may be referred to as a beginning point for further studies.

We would like to thank the editorial team for lending their expertise to make the book truly unique. They have played a crucial role in the development of this book. Without their invaluable contributions this book wouldn't have been possible. They have made vital efforts to compile up to date information on the varied aspects of this subject to make this book a valuable addition to the collection of many professionals and students.

This book was conceptualized with the vision of imparting up-to-date and integrated information in this field. To ensure the same, a matchless editorial board was set up. Every individual on the board went through rigorous rounds of assessment to prove their worth. After which they invested a large part of their time researching and compiling the most relevant data for our readers.

The editorial board has been involved in producing this book since its inception. They have spent rigorous hours researching and exploring the diverse topics which have resulted in the successful publishing of this book. They have passed on their knowledge of decades through this book. To expedite this challenging task, the publisher supported the team at every step. A small team of assistant editors was also appointed to further simplify the editing procedure and attain best results for the readers.

Apart from the editorial board, the designing team has also invested a significant amount of their time in understanding the subject and creating the most relevant covers. They scrutinized every image to scout for the most suitable representation of the subject and create an appropriate cover for the book.

The publishing team has been an ardent support to the editorial, designing and production team. Their endless efforts to recruit the best for this project, has resulted in the accomplishment of this book. They are a veteran in the field of academics and their pool of knowledge is as vast as their experience in printing. Their expertise and guidance has proved useful at every step. Their uncompromising quality standards have made this book an exceptional effort. Their encouragement from time to time has been an inspiration for everyone.

The publisher and the editorial board hope that this book will prove to be a valuable piece of knowledge for students, practitioners and scholars across the globe.

Index

www.ingramcontent.com/pod-product-compliance
Lightning Source LLC
Chambersburg PA
CBHW061956190326
41458CB00009B/2887